Innovations in Fish Passage Technology

Support for publication of this book was provided by

U.S. Geological Survey, Biological Resources Division

U.S. Army Corps of Engineers, Walla Walla District

The Bioengineering Section of the American Fisheries Society

National Marine Fisheries Service, Portland Office

Innovations in Fish Passage Technology

Mufeed Odeh
Editor

American Fisheries Society

Suggested citation formats

Chapter in the book

Amaral, S. V., and five coauthors. 1999. Fish diversion effectiveness of a modular inclined screen system. Pages 61–78 *in* M. Odeh, editor. Innovations in fish passage technology. American Fisheries Society, Bethesda, Maryland.

Entire book

Odeh, M., editor. 1999. Innovations in fish passage technology. American Fisheries Society, Bethesda, Maryland.

The new fish passage design by Mufeed Odeh, shown on the cover, was drawn by John Noreika.

Library of Congress Catalog Number: 99-65149
ISBN: 1-888569-17-4

American Fisheries Society
5410 Grosvenor Lane, Suite 110
Bethesda, Maryland 20814-2199
USA

Contents

Contributors

William Acker (Chapter 11)
BioSonics, Inc., 4027 Leary Way NW, Seattle, Washington 98127, USA

Noah S. Adams (Chapter 7)
U.S. Geological Survey, Biological Resources Division, Columbia River Research Laboratory, 5501A Cook-Underwood Road, Cook, Washington 98605, USA

Rowland Adeniyi (Chapter 6)
Hydroacoustic Technology, Inc., 715 NE Northlake Way, Seattle, Washington 98105, USA; Consulting@HTIsonar.com

Stephen V. Amaral (Chapter 4)
Alden Research Laboratory, Inc., 30 Shrewsbury Street, Holden, Massachusetts 01520, USA

Scott D. Evans (Chapter 7)
U.S. Geological Survey, Biological Resources Division, Columbia River Research Laboratory, 5501A Cook-Underwood Road, Cook, Washington 98605, USA

Julie K. Everard (Chapter 13)
Fawley Aquatic Research Laboratories Ltd., Fawley, Southampton, Hampshire, SO45 1TW, UK (Present address: Environment Agency, Riversmeet House, Newtown Industrial Estate, Northway Lane, Tewkesbury, Gloucestershire, GL20 8JG, UK)

David Fuhriman (Chapter 11)
BioSonics, Inc., 4027 Leary Way NW, Seattle, Washington 98127, USA

Brian N. Hanson (Chapter 3)
Normandeau Associates, Inc., 224 Old Ferry Road, Brattleboro, Vermont 05301, USA

John Hedgepeth (Chapter 11)
BioSonics, Inc., 4027 Leary Way NW, Seattle, Washington 98127, USA; jhedgepeth@biosonicsinc.com

Paul G. Heisey (Chapter 8)
Normandeau Associates, 1921 River Road, Post Office Box 10, Drumore, Pennsylvania 17518, USA

Tom K. Iverson (Chapters 9 and 10)
Hydroacoustic Technology, Inc., 715 NE Northlake Way, Seattle, Washington 98105, USA; consulting@HTISonar.com

Julie E. Keister (Chapter 9)
Hydroacoustic Technology, Inc., 715 NE Northlake Way, Seattle, Washington 98105, USA; consulting@HTISonar.com (Present address: Cooperative for Marine Resource Studies, 2030 S. Marine Science Drive, Newport, Oregon 97365, USA)

Joe E. Kelly (Chapter 7)
U.S. Geological Survey, Biological Resources Division, Columbia River Research Laboratory, 5501A Cook-Underwood Road, Cook, Washington 98605, USA

Daniel R. Kenney (Chapters 7 and 8)
U.S. Army Corps of Engineers, Walla Walla District, 201 North 3rd Avenue, Walla Walla, Washington 99362, USA (Present address: National Marine Fisheries Service, 2900 Stewart Parkway, Roseburg, Oregon 97470, USA)

Michel Larinier (Chapter 2)
CSP (National Council of Inland Fisheries), CEMAGREF (French Institute of Agricultural and Environmental Engineering), Institut de Mécanique des Fluides, Avenue du Professeur Camille Soula 31400, Toulouse, France

Martin Mallen-Cooper (Chapter 12)
NSW Fisheries Office of Conservation and the Cooperative Research Centre for Freshwater Ecology, Post Office Box 21, Cronulla, NSW, Australia, 2230 (Present address: Fishway Consulting Services, 8 Tudor Pl., St. Ives Chase, NSW, 2075, Australia)

Dilip Mathur (Chapter 8)
Normandeau Associates, 1921 River Road, Post Office Box 10, Drumore, Pennsylvania 17518, USA

Robert D. McDonald (Chapter 9)
Public Utility District No. 1 of Chelan County, 327 N. Wenatchee Avenue, Wenatchee, Washington 98801, USA

Thaddeus R. Mosey (Chapter 5)
Chelan County Public Utility District, Fish and Wildlife Department, 327 N. Wenatchee Ave., Wenatchee, Washington 98801, USA

Mufeed Odeh (Editor, Chapter 1)
U.S. Geological Survey, Biological Resources Division, S. O. Conte Anadromous Fish Research Center, 1 Migratory Way, Turners Falls, Massachusetts 01376, USA

Edward Paolini (Chapter 4)
Niagara Mohawk Power Corporation, 300 Erie Boulevard West, Syracuse, New York 13202, USA

Russell W. Perry (Chapter 7)
U.S. Geological Survey, Biological Resources Division, Columbia River Research Laboratory, 5501A Cook-Underwood Road, Cook, Washington 98605, USA

Charles M. Peven (Chapter 5)
Chelan County Public Utility District, Fish and Wildlife Department, 327 N. Wenatchee Ave., Wenatchee, Washington 98801, USA

Anthony Plizga (Chapter 4)
Stone & Webster Engineering Corporation, 245 Summer Street, Boston, Massachusetts 02210, USA

John M. Plumb (Chapter 7)
U.S. Geological Survey, Biological Resources Division, Columbia River Research Laboratory, 5501A Cook-Underwood Road, Cook, Washington 98605, USA

Dennis W. Rondorf (Chapter 7)
U. S. Geological Survey, Biological Resources Division, Columbia River Research Laboratory, 5501A Cook-Underwood Road, Cook, Washington 98605, USA

John R. Skalski (Chapter 8)
University of Washington, 1325 4th Avenue, Suite 1820, Seattle, Washington 98101-2509, USA

Tracey W. Steig (Chapter 6)
Hydroacoustic Technology, Inc., 715 NE Northlake Way, Seattle, Washington 98105, USA; Consulting@HTIsonar.com

Charles W. Sullivan (Chapter 4)
Electric Power Research Institute, 3412 Hillview Avenue, Palo Alto, California 94304, USA

Edward P. Taft (Chapter 4)
Alden Research Laboratory, Inc., 30 Shrewsbury Street, Holden, Massachusetts 01520, USA

François Travade (Chapter 2)
Electicité de France, Etudes et Recherches, 6 Quai Watier - 78401, Chatou CEDEX, France

Andrew W. H. Turnpenny (Chapter 13)
Fawley Aquatic Research Laboratories Ltd., Fawley, Southampton, Hampshire, SO45 1TW, UK; a.turnpenny@fawley-arl.co.uk

Frederick C. Winchell (Chapter 4)
Alden Research Laboratory, Inc., 30 Shrewsbury Street, Holden, Massachusetts 01520, USA

Foreword

A healthy environment is the key to a healthy economy, and healthy stocks of anadromous fish are essential to the nation's economy, supporting as it does important commercial, recreational, and tribal fisheries. Marine and anadromous fisheries contribute over US$25 billion annually to the U.S. economy; nearly 300,000 men and women are full-time workers in U.S. commercial fisheries; marine recreational fishing adds to the quality of life for over 17 million Americans.

As we approach the end of the millennium, it is becoming increasingly clear that the health of our valuable fisheries is dependent on the health of, and access to, anadromous fisheries habitat. Access may be prevented by small-scale blockages such as poorly placed culverts and unscreened irrigation diversions that are small individually, but have a significant cumulative impact. Blockages may also consist of large-scale dams built by private or federal entities to generate electricity, flood control, water storage for irrigation, and so on. The problems associated with the effective passage of fish around blockages to their habitats is one of the most complex and difficult challenges in the entire field of fishery management, requiring a broad interdisciplinary approach.

Hydropower represents approximately 98% of the renewable energy production in the United States. Although hydropower is cleaner than fossil fuel and nuclear power, hydropower facilities have had major adverse effects on fish populations. Although these facilities are not the only cause of fish population declines, their contributions to population declines widely recognized. Efforts to reduce environmental problems associated with hydropower operations have received considerable attention in the past decade both at federal facilities and the non-federal facilities licensed by the Federal Energy Regulatory Commission (FERC). Congress has authorized federal, state, and tribal fish management agencies to mandate fish passage and recommend fish protection measures in licensing non-federal hydropower facilities. The FERC relicensing process provides resource agencies with an opportunity to reexamine operations and further the restoration of fisheries and improvement of water quality in the nation's rivers by implementing modern environmental protections.

The Federal Power Act requires that relicensing be conducted in light of the laws and regulations in effect at the time of relicensing. Consequently, hundreds of dams licensed some 30 to 50 years ago will have to come into compliance with current laws during their relicensing. Many of these have fish passage facilities that have been less effective than anticipated and many dams have never had fish passageways during the preceding 30 to 50 year license term. Therefore, relicensing provides an unparalleled opportunity to reconfigure inadequate fishways and to prescribe fishways for many projects that have existed without them. For this reason, the application of the technological advances reported in this volume during relicensing holds great promise for significantly improving fish passage at dams and enhancing protections for aquatic ecosystems.

The objective of this volume is to give participants in the relicensing process the latest research and evaluations on the effectiveness of different fish passage techniques. However, in order for the reader to appreciate the context within which fish passage technology is developed and applied, it is necessary to describe the technical, regulatory, and political arenas within which these activities occur. In the technical context, fish passage is an intriguing engineering challenge that draws upon hydraulics, bioengineering, ichthyomechanics, and other esoteric disciplines. In the regulatory context, the application of this expertise occurs during the protracted and multifaceted five-year FERC relicensing process for non-federal dams that are guided by the Federal Power Act.

In the political context, the paucity of effective fish passage is increasingly recognized as a problem thwarting the benefits of the diverse efforts by state and federal managers to improve the quality of anadromous fish habitat on federal, state, and private lands via large-scale region wide planning initiatives, such as the Northwest Forest Plan. In light of all the efforts to improve the upland habitats of anadromous fish, it is essential that the community of hydropower stakeholders encourage use of the best possible scientific and engineering solutions to facilitate fish passage around blockages. Yet from the perspective of the anadromous fish, all the debate on regulatory reform of the relicensing process and all the interagency efforts to improve fish habitats will be of little avail if the fish cannot get there.

The problems associated with fish passage occur globally, and solutions are being explored in many countries. Although this volume primarily contains papers on research conducted in the United States on commercial species (anadromous and otherwise), the inclusion of papers from France and Australia underscores the universal nature of fish passage problems and the need to address passage for species of ecological as well as commercial value.

In part, this volume is a legacy of the American Fisheries Society (AFS) Bioengineering Section symposium at AFS's 127th annual meeting in Monterey, California. The symposium, the development of this volume, and the debate and interactions generated along the way pay tribute to the ability of AFS members to conduct fisheries research of high salience to their profession and to society at large. Mufeed Odeh deserves special recognition for his dedicated effort to make this volume a reality, and another milestone in the American Fisheries Society's tradition of excellence.

Stephen M. Waste
Office of Habitat Conservation
National Marine Fisheries Service
Silver Spring, Maryland

Acknowledgments

I am indebted to all those who helped me put this volume together. A product such as this takes a monumental effort from many. My heartfelt thanks goes to all the authors, the reviewers, my technical assistants, and the folks in the book department at the American Fisheries Society. A special thank you is reserved for my loving and ever patient wife Marilyn. The Bioengineering Section of AFS was the initial sponsor of this work. Funding for this book was provided by the U.S. Geological Survey, Biological Resources Division; U.S. Army Corps of Engineers, Walla Walla District; the Bioengineering Section; and the National Marine Fisheries Service, Portland office.

Mufeed Odeh
Conte Anadromous Fish Research Center
Biological Resources Division, U.S. Geological Survey
Turners Falls, Massachusetts

Symbols and Abbreviations

The following symbols and abbreviations may be found in this book without definition. Also undefined are standard mathematical and statistical symbols given in most dictionaries.

A	ampere
AC	alternating current
Bq	becquerel
C	coulomb
°C	degrees Celsius
cal	calorie
cd	candela
cm	centimeter
Co.	Company
Corp.	Corporation
cov	covariance
DC	direct current; District of Columbia
D	dextro (as a prefix)
d	day
d	dextrorotatory
df	degrees of freedom
dL	deciliter
E	east
E	expected value
e	base of natural logarithm (2.71828…)
e.g.	(exempli gratia) for example
eq	equivalent
et al.	(et alii) and others
etc.	et cetera
eV	electron volt
F	filial generation; Farad
°F	degrees Fahrenheit
fc	footcandle (0.0929 lx)
ft	foot (30.5 cm)
ft^3/s	cubic feet per second (0.0283 m^3/s)
g	gram
G	giga (10^9, as a prefix)
gal	gallon (3.79 L)
Gy	gray
h	hour
ha	hectare (2.47 acres)
hp	horsepower (746 W)
Hz	hertz
in	inch (2.54 cm)
Inc.	Incorporated
i.e.	(id est) that is
IU	international unit
J	joule
K	Kelvin (degrees above absolute zero)
k	kilo (10^3, as a prefix)
kg	kilogram
km	kilometer
l	levorotatory
L	levo (as a prefix)
L	liter (0.264 gal, 1.06 qt)
lb	pound (0.454 kg, 454g)
lm	lumen
log	logarithm
Ltd.	Limited
M	mega (10^6, as a prefix); molar (as a suffix or by itself)
m	meter (as a suffix or by itself); milli (10^{-3}, as a prefix)
mi	mile (1.61 km)
min	minute
mol	mole
N	normal (for chemistry); north (for geography); newton
N	sample size

NS	not significant
n	ploidy; nanno (10^{-9}, as a prefix)
o	ortho (as a chemical prefix)
oz	ounce (28.4 g)
P	probability
p	para (as a chemical prefix)
p	pico (10^{-12}, as a prefix)
Pa	pascal
pH	negative log of hydrogen ion activity
ppm	parts per million
qt	quart (0.946 L)
R	multiple correlation or regression coefficient
r	simple correlation or regression coefficient
rad	radian
S siemens	(for electrical conductance); south (for geography)
SD	standard deviation
SE	standard error
s	second
T	tesla
tris	tris(hydroxymethyl)-aminomethane (a buffer)
UK	United Kingdom
U.S.	United States (adjective)
USA	United States of America (noun)
V	volt
V, Var	variance (population)
var	variance (sample)
W	watt (for power); west (for geography)
Wb	weber
yd	yard (0.914 m, 91.4 cm)
α	probability of type I error (false rejection of null hypothesis)
β	probability of type II error (false acceptance of null hypothesis)
Ω	ohm
m	micro (10^{-6}, as a prefix)
′	minute (angular)
″	second (angular)
°	degree (temperature as a prefix, angular as a suffix)
%	per cent (per hundred)
‰	per mille (per thousand)

Mufeed Odeh

Fish Passage Innovation for Ecosystem and Fishery Restoration

1

Essential to the survival of our aquatic ecosystems is their ability to provide the fish therein with easy access to spawning, feeding, and healthy habitats. Restoration efforts are well underway throughout the world to re-establish dwindling stocks of all kinds of fish in small, medium, and large fragmented rivers. However, fragmented fish populations make fishery management less effective and more costly. Over the past three decades, fish passage technologies have been concerned with a broader array of species than in the past: anadromous (salmonid and nonsalmonid), catadromous, amphidromous, and even riverine fishes are now targeted for restoration. To be successful, innovative fish passage technologies must be biologically sound engineering design—a fish passage system is successful only when it operates as an integral system that attracts fish and passes them safely to their destination, upstream or downstream past an obstruction in the river.

This first chapter briefly explores the history of fish passage technology development, fish passage policy and legislation, and presents currently used upstream and downstream fish passage systems. Evaluation techniques used in field and laboratory studies to determine the biological effectiveness of a passage system are mentioned. The importance of multidisciplinary team approaches to fishery restoration efforts is discussed. This chapter also points out future directions of fish passage technology and new research, as well as briefly presenting the remaining chapters in the book. The intention here is to disseminate valuable information; the sharing of knowledge and information about fish passage technologies can only help in our common objective of removing barriers that hamper fish movement and fishery restoration.

Chapter 1

Introduction

Fish passage is no longer associated "only" with dams used for hydropower generation. Numerous rivers are fragmented by obstructions used for water storage, irrigation, and flood control among other uses. Also, every watershed has a unique mix of land use activities affecting fish populations, but in some watersheds the stress imposed by dams may cause the largest negative impact. The result of adverse land uses and interruption of the river continuum has been the decline and in some cases a complete loss of some fish species, including salmonid, nonsalmonid, and resident fishes (Jungwirth 1998; Peter 1998).

Effective fish passage is critical to the protection and recovery of many fish stocks. Hydropower dams and other obstructions in the river need to have effective fish passage in order to minimize their impact on the aquatic ecosystem. Fishways provide passage for healthy existing populations, passage for depleted or extirpated populations as part of a restoration program, and passage as a means of providing access to underutilized habitat areas.

Much of the publicity about the impacts of engineering projects on fish movement has come from high-profile, large-scale hydropower projects such as those on the Columbia River (the Pacific Northwest of the United States). The issue, however, is equally important to smaller hydropower plants, right down to those of less than 1 MW capacity, of which there are many more worldwide. In Europe, the drive to meet lower emission targets for greenhouse gases has fueled the quest to squeeze every last drop of hydropower potential. Low-head, run-of-river hydropower plants "process" large amounts of water and, potentially, have a higher impact on fish populations per unit of electricity generated than do larger plants. The need for sustainable development of power resources and other water uses makes it essential to solve fish passage issues at all scales.

Fish habitats downstream of a dam can be altered variously by changing water quantity, temperature, depth, velocity, dissolved gases, and sediment load. Thus, obstructions in the river may not only block access and flood spawning areas, but may also disrupt instream flow and change the pattern of gravel movement and silt deposition to conditions that are often different than the seasonal variations of the river to which fish are adapted. Injury or mortality of turbine-passed fish can become significant when fish have to pass many dams. Changes upstream of a dam are also important. Reservoirs created by dams inundate most mainstem spawning habitat, create habitat for predators, and increase the time it takes for migrants to complete their downstream journey to the ocean. Even riverine fish may need to move to other areas of the river for feeding, spawning, resting, or surviving in different seasons.

All these factors have cumulative, deleterious affects on the integrity of aquatic ecosystems, resulting in the historic and ongoing loss of fish habitat, a loss that has contributed to the extinction of numerous fish stocks and the decline of many more. For these reasons, effective fish passage systems are key to a healthy aquatic environment.

Unfortunately, relatively few dams have effective fish passage facilities. According to a 1994 study by the U.S. Department of Energy, upstream passage and protection facilities are present only at 9.5% of the 1,825 power plants regulated by the Federal Energy Regulatory Commission (FERC) and downstream passage, protection, and mitigation facilities are present only at 13% of the total (Francfort et al. 1994). Furthermore, many of the existing passage facilities can benefit from better design, more knowledge of the physical and hydrological characteristics of the site, knowledge of fish behavior, and more effective operations and maintenance programs.

Fish passage technologies have been evolving for over 300 years. Until recently, fishway design attempts have been relatively primitive in nature (Clay 1995). Denil (1909) was the first to design a fishway with an attempt to incorporate fishery science and engineering into the design. Subsequent to Denil's efforts, many studies have been conducted on the subject of fishway design and its passage effectiveness. Also, fish passage policy and its incorporation into the specifications for construction and operation of dams and hydropower plants have received serious attention from various professional groups, conservationists, and politicians (Bates 1993).

Innovative engineering designs and biological evaluation techniques, new construction materials, and a new open-minded "green" way of thinking about the importance of fish passage are leading the way to more effective fish passage. In addition, legislative bodies throughout the world are mandating that dam operators meet the resource needs by providing properly designed passage and protection measures at their facilities. This mandate has given urgency to the development of new and innovative fish passage ideas, some of which include the invention of behavioral barriers (such as acoustic signals, strobe lights, electric fields, and bubble curtains), nature-like fishways, fish-friendly hydraulic systems, and special downstream fish passage devices such as surface bypass collectors, Eicher screens, and modular inclined screens.

Overview

Fish passage technology development

Recorded attempts to build fishways have been made as early as the seventeenth century; earlier, more primitive attempts were probably made but not documented (Clay 1995). During the colonial days in North America when a settler was granted rights to a piece of land by the Crown and the land had a river running through it, the landowner had to abide by an English common law referred to then as "Right of Fishery." The landowner would have to provide compensation to their upstream abutters or means for the fish to mitigate an obstruction placed in the river. The fishways were usually "handmade" with local designs, and no downstream passage measures were considered.

The construction of a dam at Turners Falls, Massachusetts in 1795–1798, and another at Hadley Falls in 1849 on the Connecticut River in Massachusetts, resulted in the start of anadromous fish decline (e.g., Atlantic salmon *Salmo salar*, American shad *Alosa sapidissima*, and blueback herring *Alosa aestivalis*) (Rizzo 1968, 1971). Further fish decline is believed to have resulted from construction of the Enfield Dam in Connecticut in 1880, and the present dam at Holyoke, Massachusetts in 1900. In 1866, a state legislation was enacted (probably one of the first of its kind in North America) requiring the construction of devices to permit free passage of anadromous fishes over obstructions in rivers. In 1873, the first fish passage facility was constructed at Holyoke Dam (Rizzo 1968). American shad began to return to the base of Holyoke Dam in 1935 when the Enfield Dam was lowered. Subsequently, a second fish ladder was constructed at Holyoke in 1940. However, upstream fish passage at Holyoke Dam was successful only after the construction of a fish-lift system in 1955 (Rizzo 1968).

Nemenyi (1941) presented a long bibliography of over 150 studies on fishways and migratory fish behavior that were published in Europe and the United States in the late nineteenth century and the early part of the twentieth century. The bibliography by Nemenyi (1941) listed the studies and briefly discussed the contents of a selected few. These studies addressed a wide range of fish passage related subjects including biology and mechanics of fish migration, early fishways and fish elevator designs, fish passage through turbines, and fish screens. Although some of these studies showed advanced fishway designs for the time, many of them were not based on sound scientific approach (Nemenyi 1941). At the turn of the century, Denil (1909) developed the well-known Denil fishway, which he based on scientific principles, laboratory measurements, and observation of fish behavior, (Figure 1). Since then fishway design became more scientifically based and notable advances have been made.

Further study of fishway design and observation of the fish behavior continued here in the United States (e.g., McLeod and Nemenyi 1940) and in Europe (e.g., The Committee on Fish-Passes 1942). Researchers experimented with numerous baffle arrangements in steep channels (such as the Denil and modifications of it) and pool and weir fishways. Recently

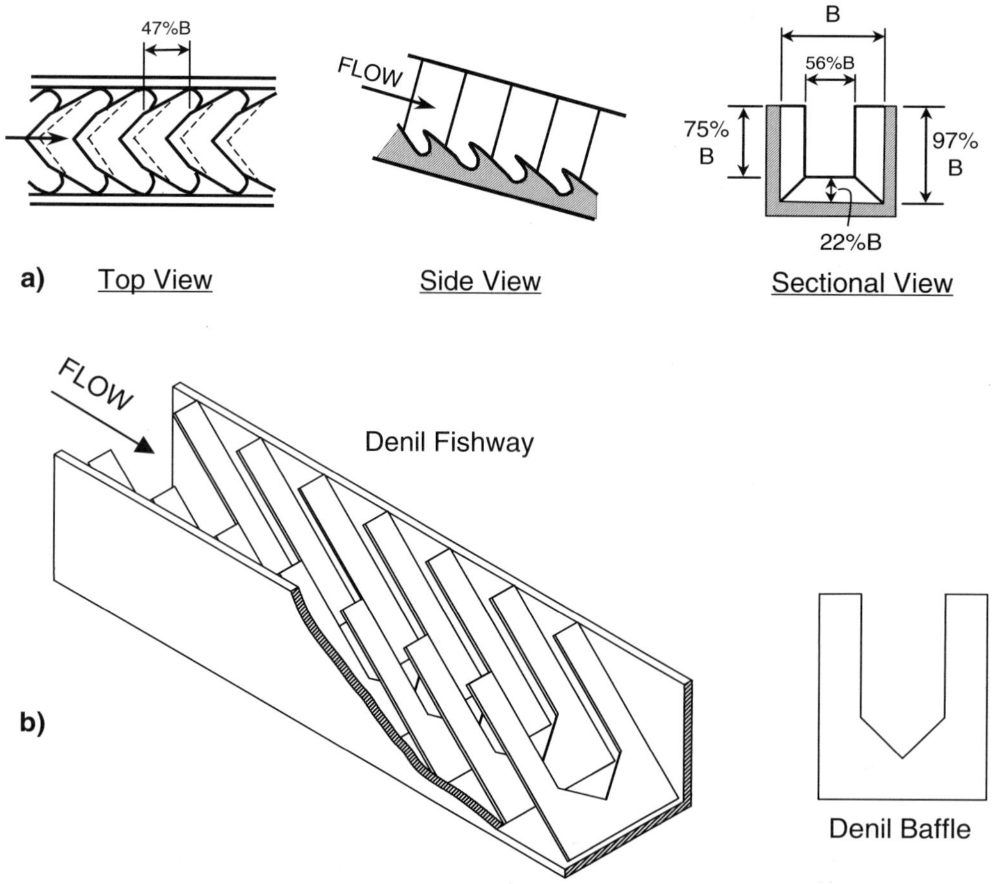

Figure 1. Denil fishway; a) original Denil design (after Denil 1909), and b) present Denil fishway, as modified by the Committee on Fish Passes (1942).

downstream fish passage has been in forefront of fish passage research and development efforts. Also, upstream passage of nonsalmonid and riverine fishes has gained popularity because of the equal importance of these fishes to the sustainability of our ecosystem.

The commercial exploitation of fish in the Pacific Northwest, land use practices, and dam building have negatively affected the Columbia and Snake Rivers salmon and steelhead since the 1800s (OTA 1995). Laws to protect the salmon, by prohibiting obstruction of access to spawning grounds, have been in place since 1848 in what was then the Oregon Territory; but were not enforced. In the early 1900s these laws were amended and rigorously enforced. Many irrigators and other water users had the desire to comply, however, working technologies were not available until 1911 when the rotating drum screen was invented (Costello 1995, cited in OTA 1995).

During the 1930s and 1940s, dams were built on the main-stem Columbia and Snake Rivers for hydropower generation. Some of the dams had pool and weir fishways incorporated into their design. Rigorous fish passage research, however, did not start until the U.S. Army Corps of Engineers built the Bonneville Fisheries Engineering Research Laboratory in 1955 near Bonneville Dam on the Columbia River. Engineers and biologists collaborated on studies dealing with several fish passage issues. Research included the rates at which fish ascended fishways, fish swimming abilities, and engineering design criteria of fishways and associated facilities. This serious commitment to research and development at that laboratory resulted in widespread generic fishway designs that are considered standards today. Research at the Bonneville laboratory continued until its demise in 1985.

Presently, mammoth efforts by biologists and engineers continue—both in the field at power plants and in the rivers—to understand fish behavior, evaluate existing passage systems, and formulate effective fish passage designs in different parts of the world (this volume; Bates 1993; Clay 1995; Jungwirth et al. 1998). In 1990, the U.S. Department of the Interior designed and constructed a unique facility dedicated solely for anadromous fish passage research on the Connecticut River in Turners Falls, Massachusetts. There, biologists and engineers collaborate on laboratory and field projects involving hydraulic design of fish passage systems, understanding fish behavior and physiology, and population dynamics of migratory fishes. Similar efforts are undertaken by fish passage enthusiasts and government entities in Europe, Japan, Australia, and Canada. Also, current development of hydropower dams in Brazil, Argentina, and China have fish passage mitigation activities going hand in hand with the engineering endeavors. Researchers, designers, resource managers, and legislators can keep in touch, know where to go for information and advice, and share successes through efforts such as creating this volume.

Fish passage policy and legislation

Fishery agencies have come to recognize the risk to migratory fishes posed by hydropower projects and other facilities that cause river fragmentation. As a result, recent years have seen the strengthening of conservation laws in many parts of the world. In the United States, the Endangered Species Act, among other laws (discussed below), has brought this issue to a head; in Europe, several countries have introduced or revised laws pertaining to fish screening (Turnpenny et al. 1998). The introduction in 1995 of new fish screening regulations in Scotland is expected to have a high impact on many small, low-head hydropower plants. Some of these cannot support operation and maintenance costs associated with the fine-meshed (~12 mm) screens required for excluding smolts from their intakes. The introduction of similar legislation in Denmark in 1994 also had the effect of rendering some small plants nonviable and led to their closure.

This type of legislation added impetus to the search for improved low- or zero-maintenance screening methods, spawning a variety of novel designs. It has also given urgency to the development of such behavioral screening methods as using acoustic signals, underwater lights and strobe lights, electric fields, louvers and bubble curtains. Government policies combined with global environmental awareness are contributing to major advancement in fish passage technologies. The policies governing the licensing of U.S. hydropower plants display these changing concerns: the manner in which fishways are dealt with in the context of generating electrical power is illustrative.

Licensing hydropower plants in the United States

In the United States, the Federal Energy Regulatory Commission (FERC) is an independent regulatory body that was created in 1977. One of the FERC's responsibilities is the licensing and inspection of private, municipal, and state hydropower projects. Under the Federal Power Act (FPA) of 1935, as amended, FERC is authorized to issue and enforce licenses, for periods up to 50 years, for the construction and continued operation of most nonfederal hydropower projects (presently estimated at 1,825; Francfort et al. 1994). In 1986 the FPA was amended by the Electric Consumers Protection Act (ECPA), strengthening the roles of fish and wildlife agencies and the nonpower values in the process of evaluating hydropower development in the United States. Under ECPA, FERC has a duty to ensure that hydropower projects are modified to achieve a better balance between power generation and protection of environmental resources.

During the relicensing process, FERC receives recommendations from fish and wildlife agencies and Indian tribes that have jurisdiction over resources that may be affected by the hydropower plant. The relicensing process (a period of five years) presents the opportunity to install passage and other fish protection measures where they do not exist, or to reconfigure them where they are ineffective. Fish and wildlife protection requirements are developed by the National Marine Fisheries Service (NMFS, within the Department of Commerce), and the Fish and Wildlife Service (FWS, of the Department of the Interior). For the nation's anadromous fish resources, the license conditions to ensure passage around hydropower facilities, as well as to provide adequate protection, mitigation, and enhancement, constitute a very critical element of the overall licensing process by FERC.

Under Section 18 of the Federal Power Act, a fishway must address the biological requirements of the upstream and downstream movement of fish, given how those requirements are affected by the structural and nonstructural elements of a hydropower project. It is important that fishways do not fail to pass fish simply because they lack the operational measures necessary to permit their use. Fishways may, therefore, be comprised of structures, facilities, devices, or operational measures necessary to ensure the safe and timely movement of fish past a hydropower project for purposes such as spawning, rearing, feeding, dispersing, and the seasonal use of habitat. The National

Marine Fisheries Service and FWS both prescribe the construction, operation, and maintenance of fishways via general directives, specific standards, or design criteria or plans, and may address such issues as site access and operational measures necessary to ensure fishway effectiveness including monitoring, evaluation, compliance, and modification.

An important section of the amended FPA is Section 10(j). Here FERC has a broad mandate to consult with federal and state fish and wildlife agencies to obtain their recommended terms and conditions. Section 10(j) allows FERC to reject or modify agency recommendations if it determines that they are inconsistent with the purposes of the requirements of the FPA or other applicable law. The Federal Energy Regulatory Commission is required to attempt to resolve conflicts over 10(j), "giving due weight to the recommendations, expertise, and statutory responsibilities of such agencies." Most disagreements are resolved informally between FERC and the agencies. If FERC does not adopt an agency's recommendations, however, it must publish the following:

- that adoption of the recommendations would be inconsistent with the purposes and requirements of the FPA or some other applicable law; and,
- that the conditions selected by FERC comply with its mandate to adequately and equitably enhance, protect, and mitigate damages to fish and wildlife (including related spawning grounds and habitat) that are affected by the development, operation, and management of the hydropower project.

Major initiatives to restructure FERC's relicensing process are underway to develop practical ways to improve the overall relicensing process. These are undertaken by groups of government and private industry participants, such as the Interagency Task Force to Improve Hydroelectric Licensing Processes (ITF) and the Electric Power Research Institute (EPRI). Although the ITF and EPRI efforts will continue into the year 2000, substantial progress has already been made to improve relicensing from the perspectives of license applicants, resource agencies, and nongovernmental organizations. Hopefully, the resulting administrative improvements to the regulatory process will contribute to a more positive political environment, which should in turn benefit the development of passage technologies, and ultimately benefit fish.

Currently Used Fish Passage Systems

Anadromous fish spawn in fresh waters and migrate to the salt waters of the oceans to live to maturity. Catadromous fish spawn in the oceans and the juveniles find their way up rivers to live their life until they are ready to spawn again. Amphidromous fish move between the ocean and the estuary during their spawning runs, without venturing too far upstream. (The term *diadromous* refers to all such migration between salt- and freshwater.) The presence of natural and man-made obstacles (such as dams) make it difficult, and sometimes impossible, for these fishes to reach their spawning and feeding habitats. As outlined above, the negative effects of dams on fish populations had been realized long ago. The result is that passage and protection measures have been in development for many years. The following is a brief discussion of currently used upstream and downstream fish passage technologies.

Upstream fish passage

The initial focus on fish passage has invariably been on restoring upstream runs of migratory species, especially salmonids and alosids: the fish must go upstream before they can come downstream. In the beginning, fishways were constructed haphazardly using available materials and without proper engineering design or understanding of fish behavior. The Denil fishway was the first to be designed based on scientific methods (Denil 1909). It has baffles on the sides and bottom of a straight channel (Figure 1a). The baffles act as energy dissipators and produce secondary flows (in the upstream direction) that reduce the water velocity in the main channel to values passable by the target upstream migrants.

The Committee on Fish Passes (1942) stated that modern fish passes (at the time) were of four main types; pool, steep-channel, fish lock, and fish lift or elevator type. Fishways today belong to the same categories (Clay 1995). As a result of their research, the Committee on Fish Passes (1942) modified the baffles of the original Denil fishway so they may be easily fabricated, resulting in today's simple Denil fishway (Figure 1b). Another form of the Denil fishway was developed in the early 1960s. The Alaska Steeppass (Figure 2), is used today for upstream migrating salmonids and alosids (Zeimer 1962; Haro et al. 1999). The Steeppass is often used in remote areas because it can be prefabricated in sections, transported, and easily installed on location.

Pool-type fishways may have weirs of different shapes and openings in them dividing the pools to provide gradual hydraulic head drop and water velocities that can easily be negotiated by the fish (Figure 3). A vertical slot fishway differs in that the weirs between pools comprise two sections forming a slot spanning the total height of the weir (Figure 4). Vertical slot fishways are often used at locations where the water level in the upstream forebay and the downstream tailrace fluctuates greatly be-

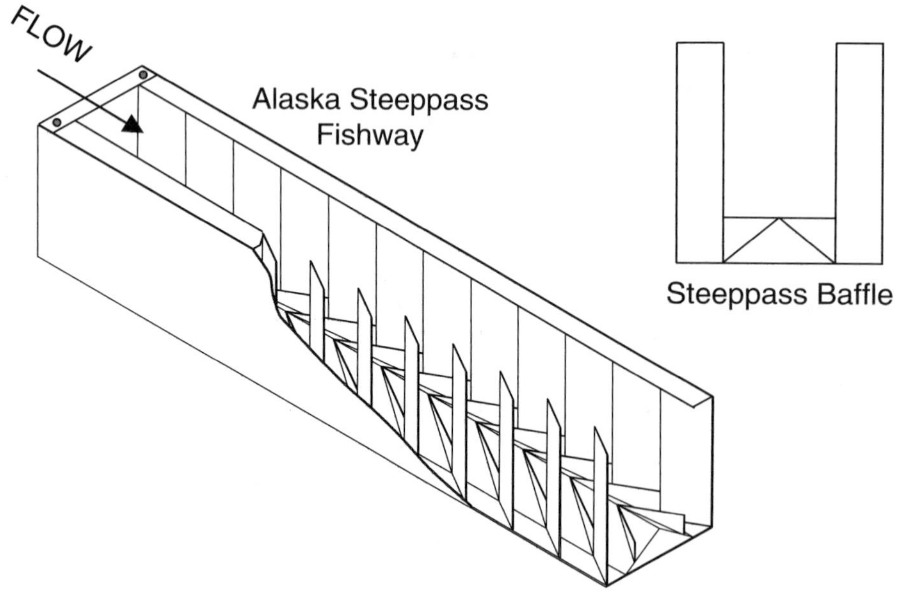

Figure 2. Alaska Steeppass fishway; another form of the Denil fishway with the baffles arranged slightly differently from the original or modified designs and the entire fishway made portable.

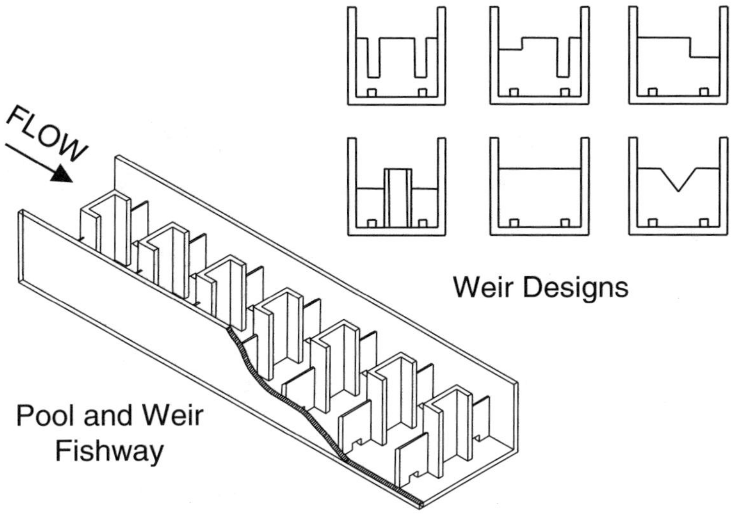

Figure 3. Pool and weir fishway showing various possible weir designs.

cause of their ability to cope with these changes. Culvert fishways are often used to provide access to migratory and riverine fishes whose route was interrupted by the construction of a roadway across a stream. Culvert fishways are basically the pipe culvert outfitted with energy dissipating baffles of various shapes and sizes to reduce the water velocity and allow the fish to swim upstream (Figure 5).

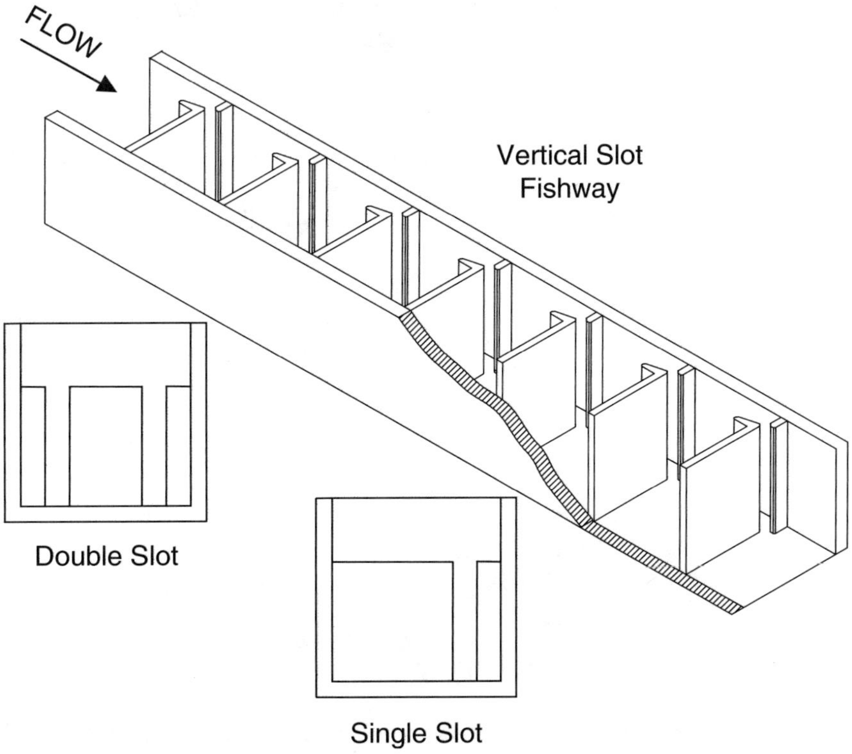

Figure 4. Vertical slot fishway showing possible one or two slots within a weir between pools.

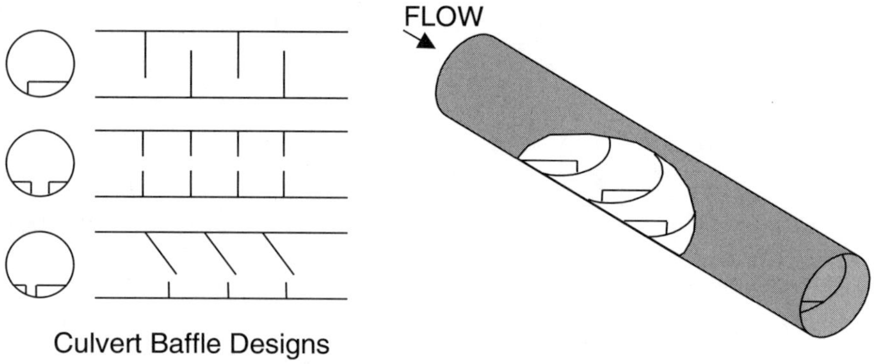

Figure 5. Culvert fishway showing various possible baffle designs.

Eels are catadromous fish; the juveniles (elvers) migrate upstream from the ocean to their habitat areas in the freshwaters of rivers. These fish are unique in that they can slither like snakes given the tiniest amount of water. Different forms of eel fishways have been used. The one shown in Figure 6 is basically a steep-channel type (sometimes a pipe is used instead) with bristles installed in its bottom. A small amount of water is supplied at the upstream end of the fishway, and elvers just work their way between the bristles up to the top.

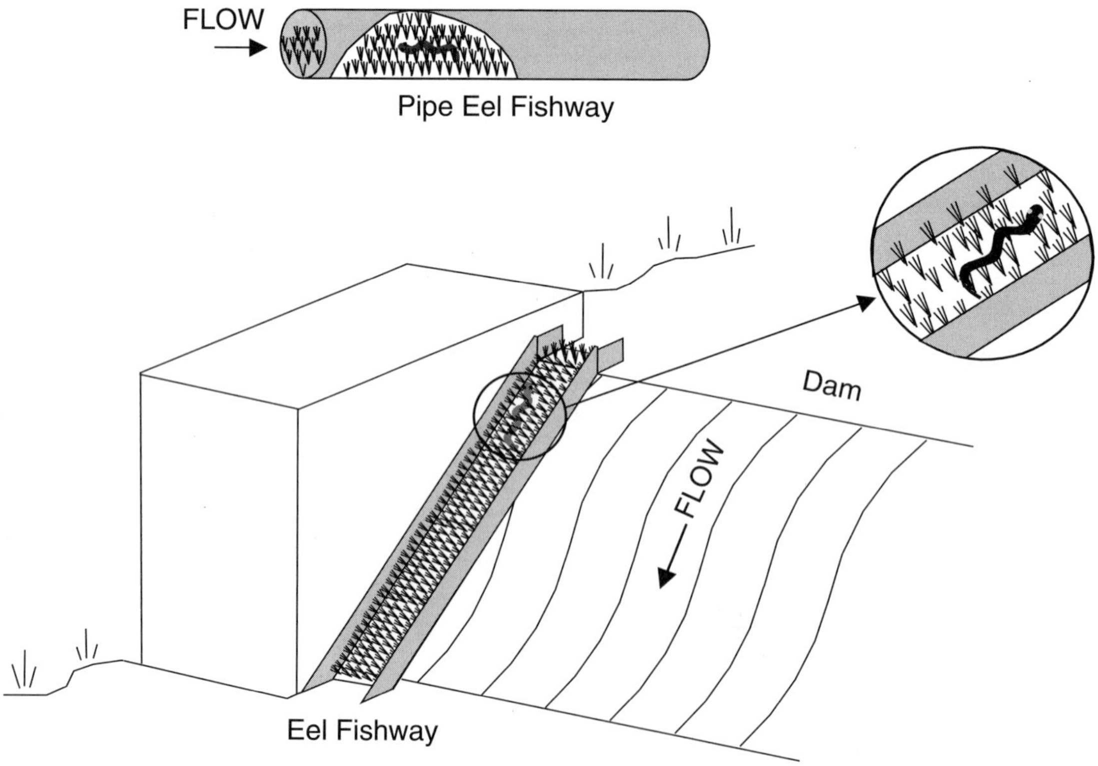

Figure 6. Eel fishways can be made out of a pipe or an open channel lined with synthetic material and placed on the downstream face of a dam.

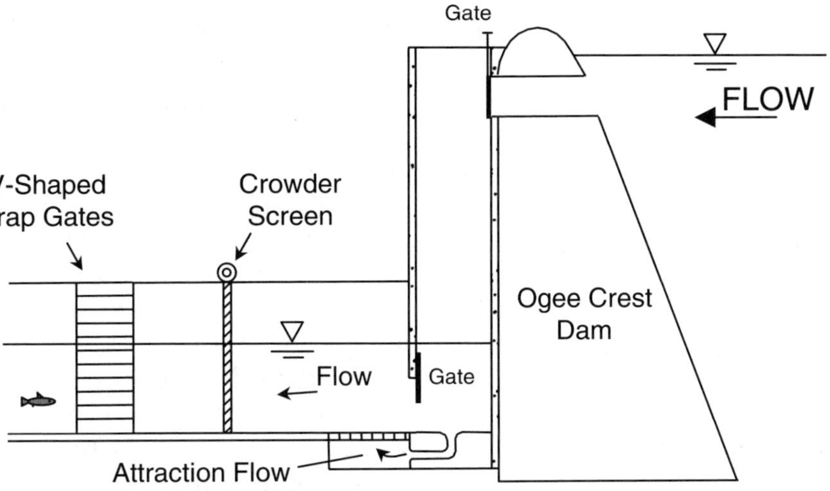

Figure 7. Schematic of a fish lock. Navigation locks are also known to allow fish to mitigate obstructions in the river.

Fish locks have been in use to assist fish in their ascent over dams since 1949 (Clay 1995). Fish enter the lock at tailwater level, a downstream gate closes, and an upstream gate opens to allow water to enter the lock; once the water level reaches the forebay elevation, the fish can exit the lock and be on their way upstream. A simple schematic of how a lock operates is given in Figure 7.

Fish elevators are used to convey fish over high-head dams. Fish volitionally enter an area downstream of the elevator's chamber, where they are crowded into the chamber (with such devices as moving screens), and the chamber is lifted to the forebay level. The fish are then dumped into the forebay or trapped and transported by tank trucks or barges to their destinations. Figure 8 shows a schematic of a fish elevator. (See Clay 1995 for examples of fish elevators from various parts of the world.)

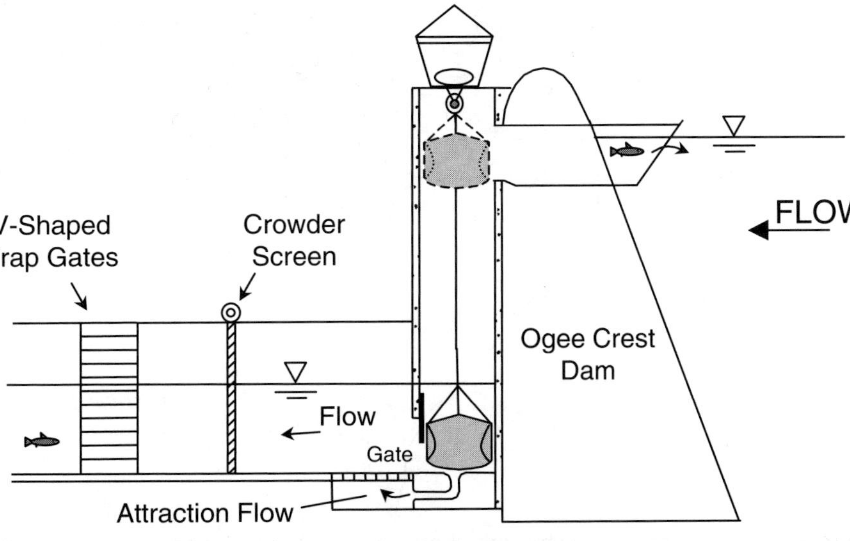

Figure 8. Schematic of a fish lift at a dam. The attraction and discharge channels at a fish lift can take on many different designs.

Recent efforts in the United States have been directed towards the outright removal of dams to provide free passage; one such is the Edwards Dam on the Kennebec River in Maine. In many cases, the choice of dam removal or installation of a conventional fishway is a costly endeavor. Removing a dam is costly at the time of removal and may cost more if the deposited sediment upstream of it was found to have toxic substances, and the operation and maintenance costs of a fishway would never stop. Some low-head dams can be breached or notched to allow fish to pass freely. The lowering of the Enfield Dam on the Connecticut River in 1935 resulted in the free passage of thousands of American shad. A notch with energy-dissipating labyrinth weirs will be placed in the crest of the Little Falls Dam on the Potomac River (starting in the summer of 1999) to open historic spawning and habitat areas to shad and herring (Figure 9; Odeh 1995).

Other methods of upstream fish passage used today include trapping and trucking, screens or bar-racks in the tailrace, navigation locks, artificial spawning channels, bypass canals, and diversion facilities.

Choices of fishway design often reflect local historical practice and ignore developments in the wider world. A recent survey of fish passage preferences in Britain (A. W. H. Turnpenny, Fawley Aquatic Research Laboratory, unpublished), for example, found strong regional preferences for particular fishway types; conflicting reasons for the choices were given. Although topography, hydrology, and biological criteria favored certain methods, it was clear that historical practice was often an overriding factor. Not one example of a vertical-slot fishway was found, despite the versatility, proven effectiveness and widespread use of this design elsewhere in Europe, North America, and Australia. The first vertical-slot fishway in the United Kingdom is presently under construction at Beeston, a small (less than 1 MW) low-head hydropower plant on the River Trent (Nottinghamshire). Here it was selected because of its ability to cope with a wide range of forebay and tailrace water levels and to pass a variety of freshwater, as well as diadromous, species. This is an example of the increasing trend in many parts of the world to open up free passage within river systems to all kinds and sizes of fish.

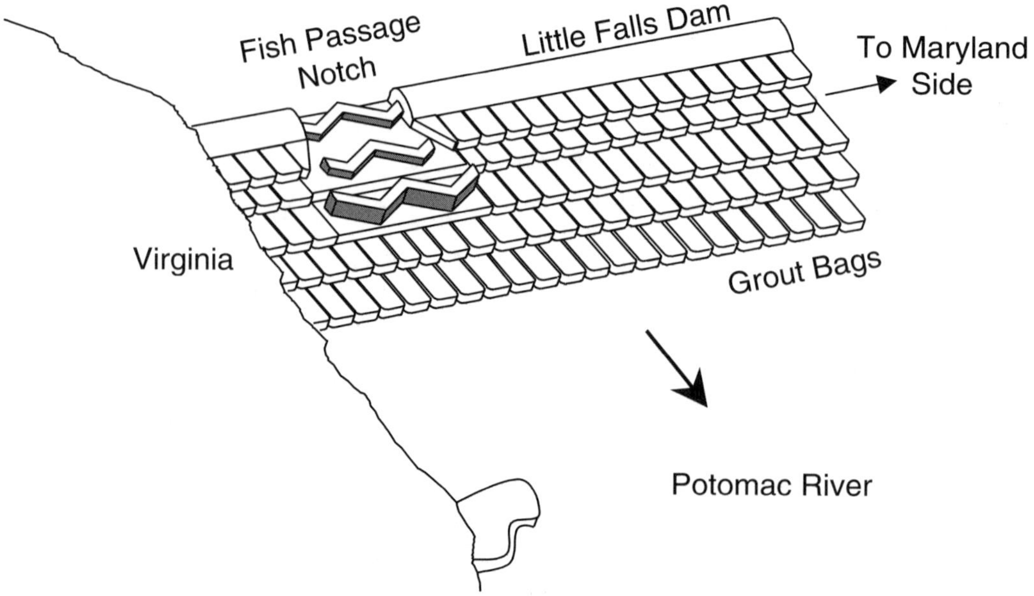

Figure 9. A fish passage notch within the top of a 3.8 m high ogee-crest dam. The notch is 7.3 m wide and 1.2 m deep. The labyrinth weirs are 0.6 m high. The largest water elevation difference is 1.2 m.

Downstream fish passage

The successful ascent of fish to upstream areas of river systems is invariably followed by the need of the same or another lifestage to move downstream. There has been much recent emphasis on this aspect of fish passage. Simple flood-control or flow gauging structures in the river may have little or no impact on the downstream movement of fish, but diversion schemes, such as for water supply and irrigation or for hydropower turbines, can lead to significant losses of down-migrating fish. Preventing fish from taking "wrong turns" into these facilities has been attempted by using various structural and behavioral barriers.

Downstream fish passage/protection measures include screens, bypasses, louvers, angled bar racks, surface spill, and behavioral means (such as light and sound). Using one or a combination of these systems at different sites yields varying results (Ruggles 1993; EPRI 1994; OTA 1995; Chapter 4 of this volume). Each site has its unique design and flow characteristics as well as fish species of interest. Therefore, careful attention is needed when designing the downstream fish passage protection system; flow approach, protection and guiding devices, and conveyance mechanism as well as the tailrace plunge pool must be designed to fit the conditions present (Odeh 1998).

Downstream fish passage systems comprise a protection/guidance device leading fish to a bypass entrance to a form of conveyance mechanism that transports the fish a short distance to a trap-and-truck facility or a plunge pool in the tailrace area. Protection and guidance devices prevent fish from entering intake structures to pumping plants or turbine penstocks. Presently used guidance/protection devices include louvers, guide walls, trash racks with reduced spacings, and trash racks overlaid with a curtain wall, perforated plate, or a reduced spacing overlay (Figure 10).

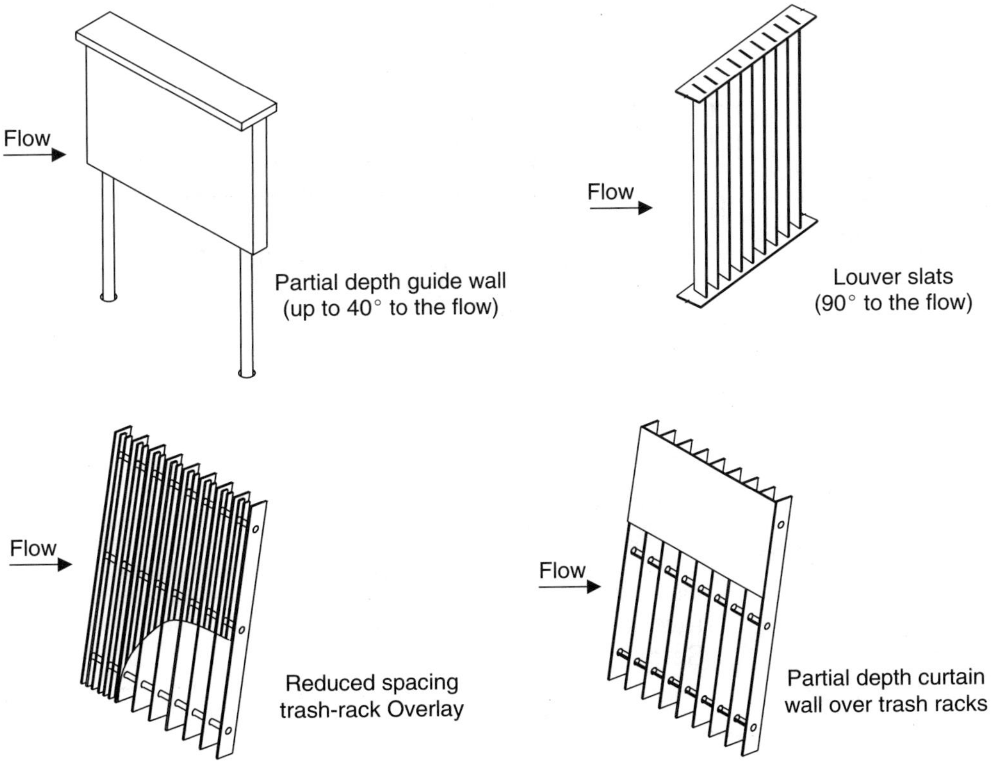

Figure 10. Examples of presently used protection/guidance mechanisms.

In some cases, conventional fine-meshed mechanical fish screens can lead to high costs as a result of not only the capital installations but also (perhaps more importantly) the loss of flow due to clogging and the high maintenance effort. This led to developments in fish behavioral barriers, such as acoustic screening (e.g., infrasound; Knudsen et al. 1997), audible sound (Lambert et al. 1997) and ultrasound (Carlson 1995) having been reported. Acoustic deterrent systems are being used increasingly in Europe, where deflection efficiencies of around 70–90% are generally being reported (Turnpenny 1999); applications include water treatment plants, irrigation schemes, thermal power plant cooling systems and hydroelectric power plants. Such behavioral methods can be used as stand-alone screening systems or in conjunction with other behavioral or mechanical methods to improve overall screening efficiency.

Bypass entrances located at the end of a guidance mechanism are of different shapes and sizes. Some are simply an opening that allows water and fish to enter the conveyance mechanism regardless of the way the entrance affects the flow hydraulics, which in some cases is believed to affect the behavior of the fish and may lead to delaying or repelling the fish rather than attracting them. Once the fish are in the system the challenge is to have them move downstream and not swim back out of it.

Conveyance/guidance devices used today to transport fish downstream of obstructions in the river are many. These include the following:

- Trapping and trucking—The fish are trapped upstream after being guided to a collection facility and hauled in tank trucks or barges to the river estuary bypassing all dams downstream.
- Submerged traveling screens (STS)—These are placed inside turbine penstocks leading fish to a vertical shaft then to a collection channel bypassing the turbines (Figure 11). The bypassed fish are then deposited into the tailrace or collected and transported to the estuary.

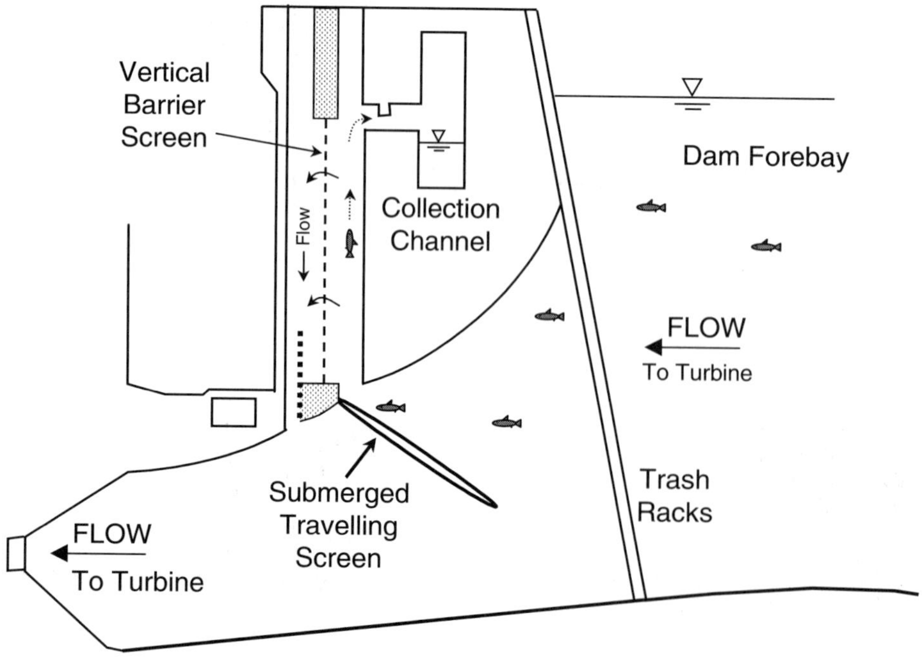

Figure 11. Schematic of a submerged traveling screen installed in a turbine penstock.

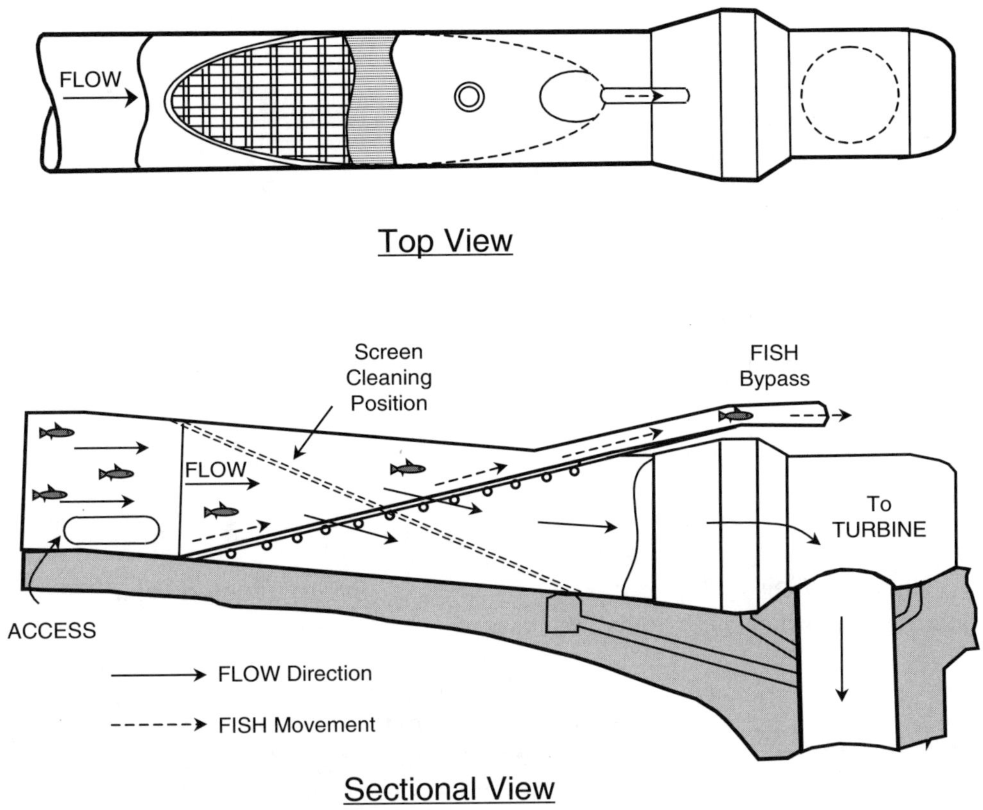

Figure 12. Eicher screen (Source: Eicher 1982).

- Eicher screen—Also called passive fish screen, this is a fixed screen placed inside the turbine penstock at about 19° to the flow (Eicher 1982). The fish are guided to an exit port at the top of the penstock bypassing the turbine, as illustrated in Figure 12.
- Specially designed weirs—These are downstream fish passage weirs in the shape of a flume (e.g., NU/ARL weir). The flume provides fish with gradually accelerating flow. Once within the zone of influence just upstream of the flume, fish were found to use it in schools of two or more compared to single fish going over a regular sharp crested weir (Haro et al. 1997). Figure 13 shows a schematic of an NU/ARL and sharp crested weirs.
- Modular inclined screen (MIS)—This device is thoroughly described in Amaral et al. in Chapter 4.
- Surface bypass collectors (SBC)—Several chapters in this volume describe the development and evaluation of these devices, see Peven and Mosey, Chapter 5.
- Spill passages—At some hydropower plants, the operation of the turbines is altered such that more water is passed over the dam and the fish are allowed to bypass the turbines.
- Trash and ice sluices—At many power plants where fish guidance/protection is not available, operators utilize the trash and ice sluice gates to provide a path for fish to bypass the turbines.

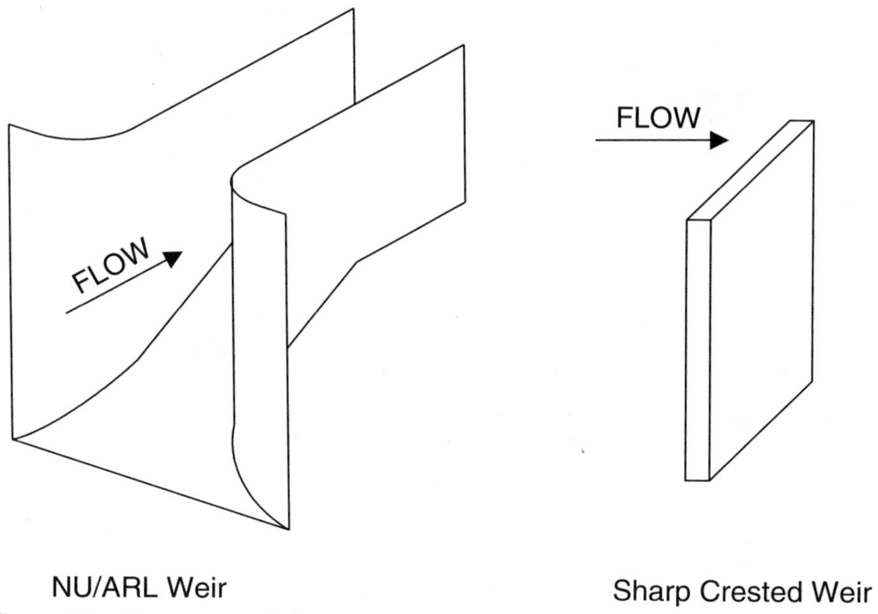

Figure 13. Downstream fish bypass weir designs.

Finally, two other issues are leading to the need to develop improved risk assessment methods for fish passage through hydroelectric turbines: the desire to avoid using mechanical or behavioral barriers and the continued use of barriers that are less than 100% effective. At the same time, these are stimulating the development of more "fish-friendly" hydraulic systems and turbine designs. A recent joint effort (starting in 1994) in the United States between the federal government (Department of Energy) and the hydropower industry was culminated by the development of fish- and environmentally friendly hydropower turbine design concepts (Odeh 1999). These concepts include a new turbine runner design and modifications to existing conventional runners. Plans to test some of these concepts are underway (Odeh 1999).

Evaluation of Fish Passage Systems

Signs of fish using a fishway are generally taken as evidence that it works, but fishery managers have been hampered by a lack of good scientific evidence about passage efficiency. The efficacy of a fish passage facility also needs to be judged within the context of others operating on the same river system. One fishway that passes 80% of fish might seem good, but five similar structures operating in series on the same river system would only allow 32% of fish to ascend to the top.

Once a fish passage facility is designed and constructed, it is important to evaluate its performance and determine its efficiency. Evaluation allows for modifications to correct faulty designs or operation, and when the design is deemed successful it can be utilized generically where physical and biological circumstances are comparable.

Significant advances and innovations in biological evaluation of performance of fish passage systems have occurred only in recent years. In the past, a passage structure such as a fishway would be deemed "successful" if sufficient numbers of fish were observed passing (Clay 1995). Quantification of passage was left largely to the observer's powers of perception of fish moving through the structure (usually from a vantage point above water), and knowledge of run size and timing. Failures of early fishway designs prompted development of systematic methods for passage design based on predictive models of fish behavior, locomotory performance, and known responses to passage structure and hydraulics (Powers and Orsborn 1985).

Analyses of passage data have also fostered a rethinking of the concepts and definition of passage performance and its quantification. For example, fish passage (both upstream and downstream) can be partitioned into three sequential modes:

- attraction (to a structure of interest, such as a fishway or downstream bypass),
- the passage itself (movement through the structure), and
- post-passage effects (stress, exhaustion, instantaneous and delayed mortality, injury, and susceptibility to predation).

Each of these modes requires different experimental and analytical techniques for its quantification, yet they may all be interdependent. Comprehensive evaluation of fish passage performance requires the ability to measure, quantify, and integrate all three modes, yet attraction and post-passage effects have received attention only recently.

Remote recording (e.g., video, resistance counters) allows for long-term, unbiased observations of fish passage, (and hence timing and patterns of movements) but only at a single point: individual fish usually cannot be identified. Recordings from underwater cameras have also revealed aspects of fish behaviors and responses to structures and hydraulics that are not entirely evident to the surface observer. Externally marking individual fish can facilitate documentation of behavior and performance of individual migrants, but also usually requires the recapture (by trapping or netting) of marked fish.

The advent of biotelemetry has made tracking of individual fish through passage structures possible, and has provided the ability to make accurate estimates of the passage of representative subsets of marked fish. Some limitations of the utility of telemetry still exist, primarily in terms of the high cost of tags (limiting sample size), minimum taggable fish size, and the potential effect of tags on fish motivation, performance, viability, and susceptibility to predation.

Marking of individuals, either with external visible tags or with telemetry, has promoted development of new statistical evaluations of passage performance. In particular, mark–recapture experiments with novel tags and techniques (Heisey et al. 1992; Mathur et al. 1996) have provided methods for estimating downstream passage mortality with increased precision. Recent investigations of downstream passage evaluation have expanded to include spillway and bypass survival (Heisey et al. 1996; Chapter 8 of this volume).

Movement and passage of small fishes (such as juvenile downstream migrants) that cannot be individually marked have been characterized and quantified using hydroacoustic monitoring techniques (Chapters 9, 10, and 11 of this volume). Hydroacoustic monitoring also offers an advantage of tracking individuals that have not been handled or marked and observing their movements through large volumes of water. Such techniques have been very useful in studies of responses of downstream migrants to hydraulic and other environmental conditions in hydropower forebays, penstocks, and bypass entrances (Skalski et al. 1993).

Another complementary approach for accurately measuring passage performance is to build and evaluate prototype passage structures in large, controlled environments, usually in the form of large flumes. Such experiments have been carried out in hydraulic laboratories or facilities designed expressly for evaluation of full-scale structures with live fish (Haro et al. 1997, 1999). The main advantage of such facilities is that the experimental prototype structures can be physically altered *in situ* on a short-term basis; such procedures usually cannot be performed with existing structures in the field, such as an operating fishway on a river. Another advantage of controlled facilities is that some external environmental variables—light, flow—can be altered to assess their effects. Combined with such tools as telemetry and high-speed and underwater video, these facilities can provide detailed information on fish behavior and response (to hydraulic and environmental variables), and on intensive tests and refinement of novel designs (e.g., Orsborn 1987; Haro et al. 1999). As a result, these types of facilities have produced a number of recent fishway and downstream bypass designs that have seen extensive, global application. In addition, passage behaviors and performance can be quantified by methods other than estimation of percent of fish passed (e.g., transit times, swimming performance, and maximum orifice velocities; Mallen-Cooper 1992).

Multidisciplinary Team Approach

Fish passage systems comprise complex hydraulic environments designed to attract and safely convey migrating fish—upstream or downstream—past obstacles on the way to their spawning, feeding, and habitat grounds. Understanding the hydraulic engineering aspects of these systems is essential. Also, knowing the behavior of the fish they "serve" (such as swimming ability and reaction towards various environmental and design conditions) increases the chances of the system being properly designed and operating more efficiently.

Successful design development of juvenile and adult fish passage/protection systems depends on efficient multidisciplinary coordination and cooperation between biologists and engineers. While purely engineering or biological schemes for addressing fisheries needs may achieve some level of success, optimum results (especially in the context of complex project locations) can only be attained when a team approach is adopted (Clay 1990; Ferguson et al. 1998; Kendall 1999).

During the last few years, fish passage teams consisting of biologists and engineers have increasingly worked together to achieve better performance of evolving fish passage technologies. This has resulted in fish passage design engineers assisting biologists in the development of innovative monitoring and evaluation techniques for prototype facilities, and biologists assisting engineers in estimating fish responses to hydraulic conditions depicted by hydraulic physical models. These teams are pushing the limit of what can be drawn from the synthesis of all available site-specific bioengineering information, and its application to refinement of new and innovative fish passage technologies.

As an example, continuing the development of fish tracking and monitoring technologies allows for a better understanding of behavioral responses of fish to hydraulic and other environmental factors within the fish passage system. These technologies include single- and multiple-beam hydroacoustics (Chapters 6, 9, and 10 of this volume), radiotelemetry methods that allow tagging of smaller fish (Chapter 7 of this volume), and passive integrated transponder (PIT) tags (Haro et al. 1999). Sonic (acoustic) tags are the latest tracking technology to be suited to juvenile migrating fish, and give a more precise three-dimensional positioning of individual juvenile routes in the forebay of large hydropower projects (Steig et al. 1998). These innovative evaluation technologies require extensive collaboration between the engineers who develop the new fish passage system designs and the biologists who understand the fish response and behavior under specific design conditions.

The computational fluid dynamics (CFD) method of solution (a three-dimensional numerical computer modeling technique) is often used by engineers to obtain detailed flow field characteristics, such as water velocities and pressures, within a hydraulic system (e.g., Sinha et al. 1998). CFD is an economical and fast way to understanding flow behavior in the hydraulic system, and compliments the use of traditional hydraulic physical models (Sinha et al. 1998). Recently, CFD mathematical models have been used to provide highly detailed and fairly accurate simulations of flow characteristics (behavior) near hydropower projects to assist in fish passage studies (Meselhe and Odgaard 1998). Like hydraulic physical models, CFD simulations enable bioengineering teams to study varying flow conditions using different structural and hydraulic designs. These simulations can be obtained in short periods of time and relatively inexpensively, but most importantly they are performed before constructing costly passage systems that may be very difficult to modify, if needed.

Superimposing data gathered from sonic tags, as well as other radiotelemetry and hydroacoustic observations, onto the output from numerical models allows more precise understanding of fish behavior and their reactions to certain hydraulic conditions. This can lead to a greater understanding of fish behavior near dams, and will provide the engineers with design criteria that can potentially lead to improved passage facilities.

The rich interaction between biologists and engineers working in tandem to optimize the design and performance of fish passage facilities is the key to successful restoration of migratory fish populations to historical levels. This interaction must be done in the early stages of a project.

Fish Passage Research and Development: Present and Future

It has become clear through the long-term failure of some fisheries that many fish passage systems constructed in previous decades have not performed as expected. While a great deal of research has gone into designing, improving and validating the internal hydraulics of various fishway designs, their failure can often be ascribed to numerous factors: poor positioning of downstream entrances, inadequate attraction flows, or failure to design for flow conditions at times of the year when fish will need to use them. In Europe, for example, intense pressure on water resources, coupled with recent unusually dry weather conditions, have meant that water levels have changed leaving tailwater levels too low and fishway entrances unreachable by fish.

Designing a fishway without proper understanding of fish behavior, and their likes and dislikes of flow characteristics and other environmental conditions, sometimes render a passage structure useless. Another recent challenge in the fish passage arena has been the desire to open up river systems to nonmigratory species; this may be regarded as the strongest influence in current fishway research and development,. Recently, attention has been given to "nature-like" fishways in Europe, Japan, Australia and here in the United States (Harris et al. 1998; Parasiewicz et al. 1998). These are desirable by biologists because they provide fish with both a passage route and sometimes a habitat for smaller fish. Also, a better understanding of attraction flow, internal fishway hydraulics, fish behavior, and providing proper environmental and operational conditions and design criteria for improving fishway performance have become high priority. Some fish passage systems are being designed to provide a combination of uses, such as a system that allows fish and boats to pass over an obstacle in the river.

New challenges are being taken on every day by engineers and biologists to help restore fish populations worldwide. New innovations in fish passage technologies, biological evaluation techniques, understanding fish behavior, and designing fish-friendly hydraulic turbine systems are being utilized to alleviate the problem of dwindling fish stocks. For remediation, it is important for all involved to understand all factors involved and to cooperate

About this Book

This book is about innovations in fish passage technology used to assist migratory fishes in their innate effort to mitigate natural and man-made obstructions. The book contains studies that were conducted to discern the biological viability, through migratory fish behavior, of innovative engineering designs constructed to give fish an alternative route to passage through hydropower turbines on their way back to the ocean.

In recent years the issue of fish passage became the focus of many workers in various professional groups, such as ecologists, engineers, hydropower developers, and resource managers. Free passage of all fish species along the entire river corridor is pivotal in the effort to regain aquatic ecosystem sustainability. Until recently, upstream fish passage had been the focus of restoration efforts of fish stocks worldwide. But we have come to realize that downstream movement of fish at dams is as essential to a migratory species survival as the success of its plight on the way upstream. This book consists of examples of newly developed downstream fish passage systems and their biological evaluation results and techniques. It provides a brief exposure to up-and-coming passage issues, such as passage of nonsalmonid fishes and the understanding of how such hydraulic phenomena as cavitation contribute to migratory fish injury and mortality.

Restoration efforts of declining migratory fish stocks worldwide can benefit from sharing information on fish passage systems design and evaluation amongst the different disciplines involved. Disseminating information in volumes like this one enables professionals to apply some aspects of the presented systems to their own. The various chapters in this book clearly demonstrate that fish passage is an area where multidisciplinary teams of professionals are needed to work on any particular project to get the job done well.

Development and evaluation of downstream fish passage systems at smaller hydropower plants in Europe and the United States are presented in Chapters 2 through 4. Larinier and Travade (Chapter 2) present several downstream passage systems and their effectiveness in France. While Hanson (Chapter 3) explores the effectiveness of a couple of downstream fish bypasses on the Connecticut River, Amaral et al. (Chapter 4) presents the innovative Modular Inclined Screen (MIS) and its effectiveness as a downstream passage system at a power plant on the Hudson River in the Northeast-United States.

Surface bypass and collection (SBC) systems on the Columbia and Snake Rivers (northwestern United States) are the latest development in downstream fish passage technology innovation. Peven and Mosey (Chapter 5) and Steig and Adenyi (Chapter 6) present the development and evaluation of the SBC at Rocky Reach Dam, respectively. Adams et al. (Chapter 7) studied the behavior of smolts approaching an SBC, while Mathur et al. (Chapter 8) estimated the mortality of smolts passing the same SBC at Lower Granite Dam on the Snake River.

The new scanning split-beam hydroacoustic technique that was used for evaluating fish passage through variable designs of surface spill gates at Rock Island Dam on the Columbia River is presented by Iverson (Chapter 10) A summary of the results of this evaluation is given in Chapter 9 by Iverson et al. Hedgepeth et al. (Chapter 11) describe fish behavior near hydropower dams using another hydroacoustic technique.

The book also contains a study on the passage of nonsalmonid fishes (Chapter 12 of this volume), which has recently become an important and challenging task to be accomplished either in the United States and abroad. Finally, understanding the relationship between hydraulic phenomena and how they injure fish has become essential to designing "fish-friendly" engineering structures. The last chapter (Chapter 13 of this volume) explores how and if cavitation can injure fish. (Cavitation is a pressure related fluid flow phenomenon that occurs in turbines and other hydraulic systems; bubbles form at vapor pressure and collapse when they reach higher pressures damaging solid surfaces.)

Although it is feasible for hydraulic engineers to manipulate the flow characteristics near obstructions that defragment the river and hinder fish migration, fish behavior can not be easily dictated by man. By observing and understanding what fish prefer in their "normative" river habitat, however, we hope to design fish passage structures that are successful in helping fish mitigate the obstructions that fragment "their" river.

Acknowledgments

My heartfelt thanks go to those who assisted me in writing this chapter. Authors of all sorts are always helped by others. I especially thank Alexander Haro, Steve Rainey, Andrew Turnpenny, and Stephen Waste for their generous contributions to this chapter. I thank John Noreika for providing me with many initial drafts of many of the figures herein. I would also like to take this opportunity to thank my best friend and wife Marilyn and my two children Suriah May and Malik for their patience and unconditional love to me.

References

Bates, K., compiler. 1993. Fish passage policy and technology. American Fisheries Society, Bioengineering Section, Bethesda, Maryland.

Carlson, T. J. 1995. Use of sound for fish protection at power production facilities: a historical perspective of the state of the art. Phase 1 final report: evaluation of the use of sound to modify the behavior of fish. Bonneville Power Administration, DOE/BP-62611-4, Portland, Oregon.

Clay, C. H. 1990. Suggestions for future research on fishways and fish facilities. Pages 1–9 *in* Proceedings of the international symposium on fishways '90 in Gifu, Japan.

Clay, C. H. 1995. Design of fishways and other fish facilities (2nd edition). Lewis Publishers, Ann Arbor, Michigan.

Committee on Fish-Passes. 1942. Report of the Committee on Fish-Passes, the Institution Research Committee, Institution of Civil Engineers. William Clowes and Sons, Limited. London.

Costello, R. J. 1995. A historical perspective and information for activities and actions affecting the Pacific salmon species, relative to development and management of land and water resources within the Columbia River basin, during the period 1792–1967. Bonneville Power Administration, Portland, Oregon.

Denil, G. 1909. Les echelles a poissons et leur application aux barrages de Meuse et d'Ourthe, Annales des Travaux Publics de Belgique.

Eicher, G. J. 1982. A passive fish screen for hydropower turbines. American Society of Civil Engineers Hydraulics Division Symposium, Jackson, Mississippi.

EPRI (Electric Power Research Institute). 1994. Fish protection/passage technologies evaluated by EPRI and guidelines for their applications. Electric Power Research Institute, EPRI Report TR-104120, Palo Alto, California.

Ferguson, J. W., T. P. Poe, and T. J. Carlson. 1998. Surface-oriented bypass systems for juvenile salmonids on the Columbia River, USA. Pages 281–299 *in* Jungwirth et al. (1998).

Francfort, J. E., and seven coauthors. 1994. Environmental mitigation at hydropower projects. Volume II. Benefits and costs of fish passage and protection. A report prepared for the U.S. Department of Energy, Idaho Falls, Idaho.

Haro, A., M. Odeh, J. F. Noreika, and T. Castro-Santos. 1997. Effect of water acceleration on downstream migratory behavior and passage of Atlantic salmon and juvenile American shad at surface bypasses. Transactions of the American Fisheries Society 127:118–127.

Haro, A., M. Odeh, T. Castro-Santos, and J. F. Noreika. 1999. Effect of slope and headpond on passage of American shad and blueback herring through simple Denil and Deepened Alaska steepass fishways. North American Journal of Fisheries Management 19:51–58.

Harris, J. H., G. Thorncraft, and P. Wem. 1998. Evaluation of rock-ramp fishways in Australia. Pages 331–347 *in* Jungwirth et al. (1998).

Heisey, P. G., D. Mathur, and T. Rineer. 1992. A reliable tag–recapture technique for estimating turbine passage survival: application to young-of-the-year American shad (*Alosa sapidissima*). Canadian Journal of Fisheries and Aquatic Sciences 49:1826–1834.

Heisey, P. G., D. Mathur, and E. T. Huston. 1996. Passing fish safely: a closer look at turbine vs. spillway survival. Hydro Review 15:2–6.

Jungwirth, M. 1998. River continuum and fish migration – going beyond the longitudinal river corridor in understanding ecological integrity. Pages 19–32 *in* Jungwirth et al. (1998).

Jungwirth, M., S. Schmutz, and S. Weiss, editors. 1998. Fish migration and fish bypasses. Fishing News Books, Oxford, UK.

Kendall, R. L. 1999. Searching for the center. Fisheries 24(1):4.

Knudsen, F. R., C. B. Schreck, S. M. Knapp, P. S. Enger, and O. Sand. 1997. Infrasound produces flight and avoidance responses in Pacific juvenile salmonids. Journal of Fish Biology 51:824–829.

Lambert, D. R., A. W. H. Turnpenny, and J. R. Nedwell. 1997. The use of acoustic fish deflection systems at hydro stations. Hydropower and Dams 1(1997):54–56.

Mallen-Cooper, M. 1992. Swimming ability of juvenile Australian bass *Macquaria novemaculeata*, (Steindachner), and juvenile barramundi, *Lates calcarifer* (Bloch) in an experimental vertical-slot fishway. Australian Journal of Marine and Freshwater Research 43:823–834.

Mathur, D., P. G. Heisey, E. T. Euston, J. R. Skalski, and S. Hays. 1996. Turbine passage survival estimation for chinook salmon smolts (*Oncorhynchus tshawytscha*) at a large dam on the Columbia River. Canadian Journal of Fisheries and Aquatic Sciences 53:542–549.

McLeod, A. M., and P. Nemenyi. 1940. An investigation of fishways. University of Iowa, Iowa Institute of Hydraulic Research, Iowa City.

Meselhe, E. A., and A. J. Odgaard. 1998. 3D numerical flow model for fish diversion studies at Wanapum dam. American Society of Civil Engineers, Journal of Hydraulic Engineering 124(12):1203–1214.

Nemenyi, P. 1941. An annotated bibliography of fishways. The State University of Iowa, Report 389, Iowa City.

Odeh, M. 1999. A summary of environmentally friendly turbine design concepts. Prepared for the U.S. Department of Energy, Report DOE/ID/13741, Idaho Operations Office, Idaho Falls.

Odeh, M. 1998. Downstream fish passage design considerations and development at hydrolelectric projects in the North-east USA. Pages 267–280 *in* Jungwirth et al. (1998).

Odeh, M. 1995. Using surface notches for fish passage at low head dams. Pages 181–187 *in* Proceedings of the international symposium on fishways '95, Gifu, Japan.

Orsborn, J. F. 1987. Fishways – historical assessment of design practices. Pages 122–130 *in* M. J. Dadswell, and five coeditors. Common strategies of anadromous and catadromous fishes. American Fisheries Society, Symposium 1, Bethesda, Maryland.

OTA (Office of Technology Assessment). 1995. Fish passage technologies: protection at hydropower facilities. Office of Technology Assessment, OTA-ENV-641, U.S. Government Printing Office, Washington, D.C.

Parasiewicz, P., J. Eberstaller, S. Weiss, and S. Schmutz. 1998. Conceptual guidelines for nature-like bypass channels. Pages 348–362 *in* Jungwirth et al. (1998).

Peter, A. 1998. Interruption of the river continuum by barriers and the consequences for migratory fish. Pages 99–112 *in* Jungwirth et al. (1998).

Powers, P. D., and J. F. Orsborn. 1985. Analysis of barriers to upstream fish migration. Report to the U.S. Army Corps of Engineers, Project 82-14, Pullman, Washington.

Rizzo, B. 1968. Fish passage facilities design parameters for Connecticut River dams: Holyoke Dam. Report to the Technical Committee for Fisheries Management of the Connecticut River Basin, Bureau of Sport Fisheries and Wildlife, Boston.

Rizzo, B. 1971. Fish passage facilities design parameters for Connecticut River dams: Turners Falls Dam. Report to the Technical Committee for Fisheries Management of the Connecticut River Basin, Bureau of Sport Fisheries and Wildlife, Boston.

Ruggles, C. P. 1993. What's new in downstream fish passage. Pages 402–416 *in* D. Mills, editor. Salmon in the sea and new enhancement strategies. Fishing News Books, Oxford, UK.

Sinha, S. K., F. Stiropoulos, and A. J. Odgaard. 1998. Three-dimensional numerical model for flow through natural rivers. American Society of Civil Engineers, Journal of Hydraulic Engineering 124(1):13–24.

Skalski, J. R., A. Hoffman, B. H. Ransom, and T. W. Steig. 1993. Fixed-location hydroacoustic monitoring designs for estimating fish passage using stratified random and systematic sampling. Canadian Journal of Fisheries and Aquatic Sciences 50:1208–1221.

Steig, T. W., R. Adeniyi, T. K. Iverson, and T. C. Torkelson. 1998. Using acoustic tags for monitoring fine scale migration routes of juvenile salmonids in the forebay of Rocky Reach dam. Report prepared for Chelan County Public Utility District No. 1, Seattle.

Turnpenny, A. W. H. 1999. The use of acoustic sources in Britain for guidance of downstream migrants. Pages 119–122 *in* R. Kamula and A. Laine, editors. Proceedings of the Nordic conference on fish passage. Direktoratet for Naturforvaltning, Trondheim, Norway.

Turnpenny, A. W. H., K. P. Hanson, and G. Struthers. 1998. A U.K. guide to intake screening legislation and best practice. ETSU (Energy Technology Support Unit), Contractors Report ETSU H/06/00052/00/00. ETSU, Harwell, UK.

Zeimer, G. L. 1962. Steeppass fishway development, Alaska Department of Fish and Game Information Leaflet 12, Juneau.

Michel Larinier and François Travade

The Development and Evaluation of Downstream Bypasses for Juvenile Salmonids at Small Hydroelectric Plants in France

2

Experiments were conducted from 1992 to 1996 at four small-scale hydroelectric plants on salmon rivers in the southwest of France to relate downstream bypass efficiency to hydraulic conditions and to the behavior of salmon *Salmo salar* and sea trout *S. trutta* smolts in the intake canal.

The maximum turbine discharge varied from 20 to 85 m^3/s and the width of intakes varied from 11 to 30 m depending on the plant. The surface bypasses were located laterally along the intake at one end of the trashrack. The mean bypass discharges varied from 0.4 to 4 m^3/s, or an average of 2 to 8% of the turbine discharge. The efficiency of the devices was evaluated by the mark–recapture technique. Radio telemetry was used to monitor movement patterns of salmon and sea trout smolts in front of the intake and near bypass entrances. Depending on the site, the mean bypass efficiency was found to be between 17 and 80%. Behavior of fish in the vicinity of the trashrack and the bypass seemed to be largely influenced by the flow pattern. Poor hydraulic conditions (turbulence, strong acceleration, upwellings) and insufficient discharge were identified through direct and video observations as being responsible for many aborted passages at the bypass entrances.

The results suggest that siting of surface bypass systems must take into account flow patterns in both the trashrack area and intake canal. It is suggested that surface bypasses associated with existing trashracks may be an acceptable mitigation technology at small-scale hydroelectric projects where it is not necessary to guarantee a highly efficient downstream passage protection.

Introduction

Over the past 15–20 years, several programs have been launched on French rivers to restore and protect migratory fish populations, mainly salmon *Salmo salar*, sea trout *S. trutta*, shad *Alosa alosa*, and lamprey *Petromyzon marinus*. In the beginning, these programs involved building upstream fish passage facilities. At the time, priority was placed on providing free movement for adult fish returning to the river. More recently, attention has been paid to problems associated with downstream migration of juveniles.

The main obstacles to downstream migration in French rivers are small-scale hydroelectric power plants. Typically, the head at the power plant is lower than 5 m and powerhouses are equipped with Kaplan turbines, causing low to moderate damage (2–10% mortality rate). Attempts are now being made to retrofit existing facilities to reduce these damages. The systematic installation of physical barriers, such as fine-mesh screens (Ruggles 1980; EPRI 1986; Clay 1995), was, however, considered to be unrealistic, as this would have required resizing most of the water intakes to obtain sufficiently uniform and low-approach velocities. Thus, research has focused on less expensive and cumbersome technology. An attractive alternative to fish screens, surface bypasses take advantage of the reluctance of fish to cross conventional intake trashracks (Ruggles 1992; EPRI 1994). Initial promising experiments were conducted at the Halsou power plant and at the Poutès dam (Bomassi and Travade 1985; Larinier and Boyer-Bernard 1991a, 1991b); a research program was then undertaken to assess the efficiency of such bypasses and to optimize their positioning and sizing, with the long-term objective of defining optimum design criteria and determining limits to their use. The program essentially involves close study of the surface bypass devices in situ. This report gives the results obtained from 1992 to 1997 on different bypasses constructed at four hydroelectric plants in southwest France: Soeix, Bedous, St. Cricq, and Camon.

Surface Bypass Evaluation Techniques

Mark–recapture method

Downstream bypass efficiency was assessed using a mark–recapture method. At each site, the study covered a period of approximately two months during peak downstream migration of salmonid juveniles; that is, March to May. The marked fish were either hatchery-reared salmon smolts or wild salmon and sea trout smolts captured upstream from the site. Hatchery-reared smolts were included to ensure adequate samples as trapping of wild salmon was uncertain and depended on their limited natural reproduction and on upstream restocking efforts that vary from site to site and from year to year. Generally, salmon smolts were found to vary in size from 14 to 24 cm (average: 18 cm) and to weigh 50 g on average; sea trout smolts were observed to be significantly bigger (20–30 cm).

The fish were marked by fin clipping and by injecting dye (alcyan blue) into fins using a jet inoculator. They were released upstream from the experiments sites in batches of 50–200 individuals (10–30 batches per study, distributed over the observation period).

An inclined screen (15 to 30% slope) system for trapping fish was installed downstream from the bypass. The space between bars was 1 cm, and porosity was around 50%. A PVC (polyvinyl chloride) plastic trough at the downstream edge of the screen collected and guided fish toward a holding pool. Depending on the site, fish were removed from the trap 3–4 times a day (at 0800, 1400, 2000, and 2400 hours). Bypass efficiency was evaluated as the relative proportion of tagged fish recaptured in the trap.

Radiotracking

During each of the studies, displacement and behavior of smolts in the approach channel near the trashracks and the vicinity of the bypass entrance was monitored by radiotracking. Tagged fish were released upstream from the plants and their trajectories monitored by calibrated recorders in predefined areas. Manual tracking complemented automatic recordings. All of the fish were released upstream from the plants at distances of 0.4 km to 2 km from the intakes.

American equipment from the Advanced Telemetry Systems company was used. Transmitters were introduced into the fish stomach using a technique similar to that used in France for adult salmon and shad (Travade et al. 1989) and elsewhere for salmonid juveniles (Moser et al. 1990; Armstrong and Rawlings 1993). The transmitters (frequency band: 48–49 MHz) weighted from 1.5 to 1.9 g and were 19.6 mm in length and 6 mm in thickness, with a lifetime of from 11 to 15 d. Fish displacement was monitored with both graphic and digital recorders that, at regular intervals of a few seconds, detected and recorded the presence of fish in predefined areas by means of underwater antennas or aerial antennas (directional loops). The detection zones were defined to identify the bank along which specific fish arrive at the plant, their displacement in the vicinity of trashracks, their passage into the zone of influence of the bypass as well as their passage through the turbines. For manual tracking of fish, we used aerial directional antennas (loops) and one underwater coaxial cable loop (12 to 16 cm in perimeter) for more precise location.

For practical reasons, a relatively small number of fish (11–36 individuals per study) were used in the radiotracking operations. This was because we could only monitor one or two fish at a time in order to determine their position at intervals of only a few seconds (as fish may move across trashracks and through bypasses very quickly). However, while these numbers were sufficient for obtaining information on fish behavior, they did not allow us to reliably determine the efficiency of the bypass.

Environmental and plant operation parameters

The upstream water level was recorded to enable evaluation of the discharge through bypasses that did not have automatic weir crest adjustment of the bypass gate. In addition, various environmental and plant operation parameters likely to influence the behavior of migrants and the efficiency of the bypasses were recorded with automatic data collection stations (every 15 min). These were the turbine discharge at each unit, spillage, water temperature, conductivity, atmospheric pressure, solar radiation, and turbidity.

Flow characteristics

Flow patterns in the intake canals and near the entrance to the bypasses were observed and countercurrents were identified through the accumulation of debris. Floats and dye (fluorescein) were also used to visualize currents, and velocity was measured with current meters, in particular immediately upstream from trashracks.

Visual observation of fish and video counting

The small size of sites allowed numerous visual observations of fish behavior. In addition, fish behavior at the entrance to bypasses was observed at the Bedous plant with the aid of an underwater video camera.

Studies of the effect of light on fish passage necessitated the continuous counting of fish using the bypass. Fish were counted by means of video cameras placed at the bypass entrance or over the trapping screens, and the data were recorded on time-lapse VCRs.

Use of light to increase bypass efficiency

A number of previous experiments (EPRI 1986) seemed to indicate that lighting the bypass at night influenced passage rates and that intermittent lighting was preferable to continuous lighting, fish passage typically occurring at the moment the light was turned on or off (Taft 1988; Larinier and Boyer-Bernard 1991b). Various tests were thus carried out on the attractive effects of lighting. Fish attraction to the bypasses was tested using a 50-W mercury vapor lamp installed about 1–1.5 m over the water surface upstream from the bypass. Four- to eight-h periods were allotted to lighting experiments. During these "lighting" periods, various frequencies and duration of lighting were tested. For most of the experiments, a 10-min lighting phase followed by 5-min dark phase was adopted.

Soeix Experiments

Installation description

The Soeix plant is located on the Aspe River, which has a mean annual discharge on the order of 24 m^3/s at the study site. Mean monthly discharges for March, April, and May are around 33, 43, and 54 m^3/s, respectively. The Soeix hydroelectric installation consists of a gravity dam that supplies the plant through a 450-m intake canal; the plant has one vertical Kaplan wheel (nominal discharge of 24.5 m^3/s) and one propeller wheel (nominal discharge of 10.3 m^3/s). The width of the intake canal varies from 6 to 10 m, and its depth is approximately 3.5 m. The water intake (Figure 1) at the plant is 15 m wide and is fitted with a trashrack (bars spaced 3.5 cm apart) with a submerged cross-section of about 50 m^2.

Flow patterns

Flow velocities in the canal are high (1.8–2 m/s at maximum turbine discharge) and their distribution throughout the sections of the canal are relatively uniform. In the vicinity of the intake, the flow becomes asymmetrical due to broadening and change in direction of the intake canal. Flow is characterized by a tangential current from the right to the left bank and

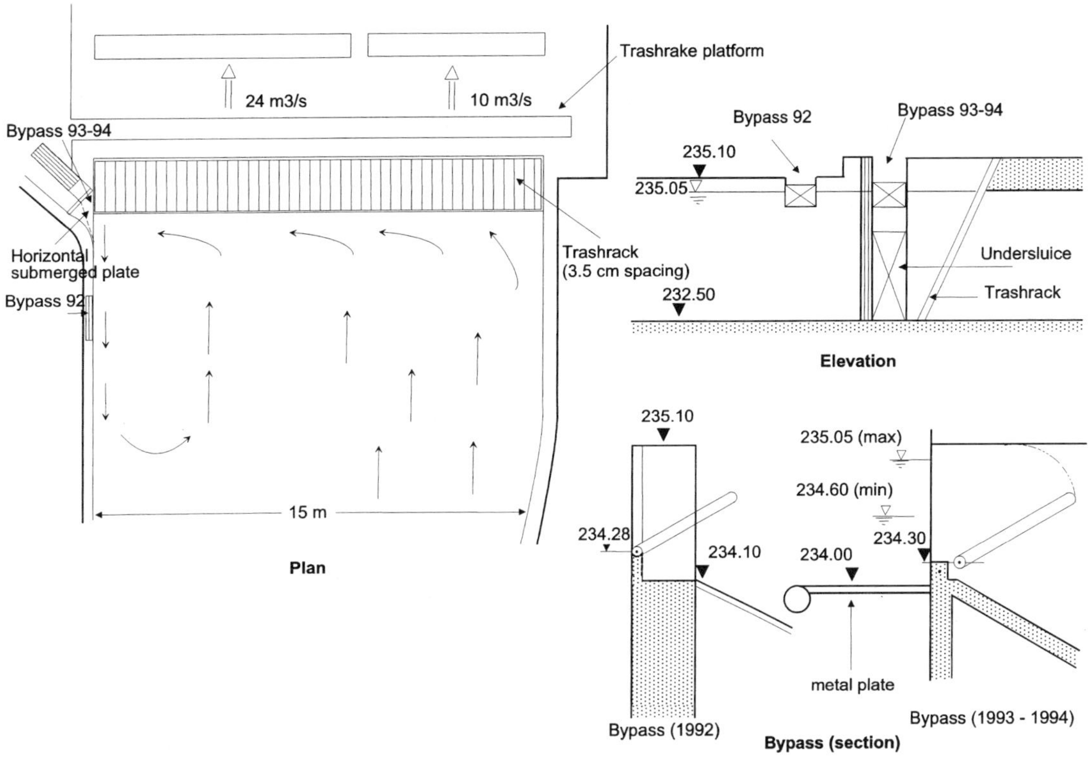

Figure 1. Schematic diagram of Soeix intake (Aspe River) and fish downstream bypasses.

by the presence on the left bank of a recirculation area covering some 30 m in length and 2–3 m in width upstream of the trashrack. Upwellings in this area were highlighted by fluorescein. In this flow pattern, spherical floats released in the headrace concentrated rapidly on the left bank, rolling along the trashrack. With both turbines operating (discharge around 28 m^3/s), the mean velocity measured just below the surface in front of the trashracks (between 0 and 0.5 m in depth) was between 0.3 and 0.9 m/s, with the lowest velocity being on the left bank.

Bypass characteristics

Three configurations of downstream migration bypasses (Figure 1) were successively set up and tested (Larinier and Travade 1996). In 1992, the bypass was on the left bank of the headrace, 6 m upstream from the screens. It consisted of a 1 × 1-m opening fitted with a flap gate manually operated by means of a pulley. Mean discharge through the bypass was 0.4 m^3/s, varying from 0.2 m^3/s to a maximum of 0.5 m^3/s. In 1993, the trash gate on the left bank immediately upstream from the screens (1.5 m) was turned into a bypass by dividing it into two sections each 0.9 m wide, only one of which was used. The mean discharge remained close to the 1992 value of 0.4 m^3/s, varying from 0.2 m^3/s to a maximum of 0.7 m^3/s. In 1994, the discharge was increased significantly (mean of 1.2 m^3/s) by opening both sections of the trash flap gate (1.8 m width). It varied between 0.5 and 1.9 m^3/s. A submerged horizontal metal plate was installed upstream from the bypass, 25 cm below the hinge axis of the trash flap gate, so as to limit the effect of upwellings and to increase surface velocities near the bypass entrance.

Flow in the immediate vicinity of the bypasses was characterized in 1992 and 1993 by upwellings that resulted from the overall pattern of currents in the approach canal. The bypass influence zone, that is, the area within which the hydraulic conditions tended to draw fish toward the bypass, was very small (less than 1 m upstream from the bypass). In 1994, under the combined effect of increased discharge and the metal plate deflector, the surface flow oriented towards the bypass was perceptible 2–3 m upstream from the bypass. The velocities varied from 0.5 to 1 m/s upstream from the horizontal plate, 0.8 to 1.4 m/s at the bypass entrance and 2.0 to 2.3 m/s over the flap gate itself.

Passage tests

Over the three years of the study, more than 3,500 fish were marked. In 1992, 1,088 hatchery-reared salmon smolts (in nine batches) and 121 wild smolts (in four batches) were released. The mean rate of recapture in the bypass was 22% (from 12.5 to 48% depending on the batch) for hatchery-reared smolts and 22% (from 19.4 to 23.3% depending on the batch) for wild smolts. In 1993, 305 wild salmon and 221 sea trout smolts were released. The mean rate of recapture in the bypass was 32% (from 13 to 39% depending on the batch) for salmon smolts and 35% (from 6 to 41% depending on the batch) for sea trout smolts. In 1994, 1,536 hatchery-reared salmon smolts (in 12 batches), 88 wild smolts (in four batches), and 149 sea trout smolts (in three batches) were released. The mean rate of recapture was 55% for hatchery-reared smolts (from 27 to 89% depending on the batch), 58% for wild smolts (46 to 77%), and 68% for sea trout smolts (65 to 72%).

Given the great number of factors influencing the test results, and their natural variability, it is only possible to give a relatively broad range for bypass efficiency, which was estimated at 20–35% in 1992, 25–40% in 1993, and 50–80% in 1994.

A total of 100 fish (86 salmon and 14 sea trout) were tagged with radio transmitters during the three years. For salmon, these were essentially hatchery-reared individuals in 1992 and wild individuals in 1993 and 1994. In 1992, 27 hatchery-reared salmon smolts and 3 wild salmon smolts were tagged, with 30% recaptured in the bypass, while in 1993, 32 wild salmon smolts and 2 sea trout smolts were tagged, with 28% recaptured in the bypass. In 1994, 24 wild salmon smolts, 2 hatchery-reared smolts, and 12 sea trout smolts were tagged, with 76% of the salmon smolts and 82% of the sea trout smolts recaptured in the bypass. It appears that the percentage of radiotracked fish going through the three bypasses was on the same order as that measured by mark–recapture methods in corresponding years.

The principal observations with respect to fish behavior on the study site are as follows:

- Migration activity was essentially nocturnal, and almost always occurred between 1800 hours and 0800 hours, though some fish did move around the study site during the day.
- The displacement of fish upstream from the study site and in the intake canal was relatively continuous and at a velocity close to that of the flow. Most salmon and sea trout (90%) paused in front of the trashrack. This points out the deterrent

effect of the trashrack with the present bar spacing (3.5 cm), but tends to indicate that the efficiency of the bypass principle will always be limited by the proportion of fish (in this case 10%) which show no reluctance to pass through the trashrack.

- In general, fish arrived in front of the screens from the left bank and most stayed there apparently due to the general flow pattern in the vicinity of the trashrack.
- Almost all fish movements remained close to the trashracks (within at most a few dozen meters). A small proportion (two salmon and three sea trout) moved more than 450 m upstream to the reservoir after reaching the trashracks. Some even went back and forth several times between the power intake and the reservoir. Most displacement of this kind occurred when the turbine discharge was between 20 and 25 m^3/s.

Lighting effect

No significant difference was observed between the rate of fish passage with intermittent lighting by the mercury vapor lamp and that with no lighting at all. Fish reacted clearly and systematically to a change in lighting, however, particularly when the light was switched off. In fact, it appeared that lighting concentrated, but did not globally enhance, bypass passages (Figure 2).

The attraction of light could not be demonstrated with radiotracked fish, as behavior differed widely from fish to fish. The effects of hydraulic attraction may have masked those of lighting.

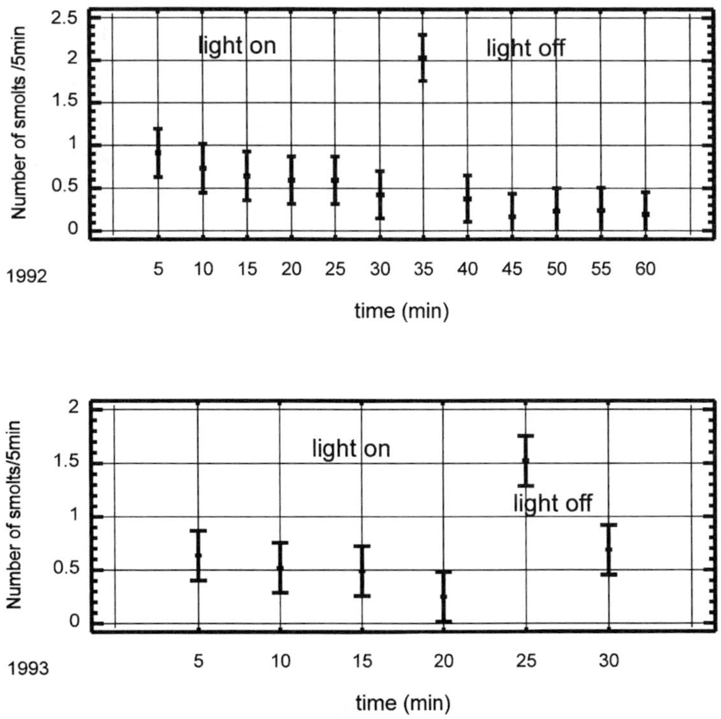

Figure 2. Rate of smolt passage under different light conditions (1992 and 1993) at Soeix downstream bypass.

Chapter 2

Bedous Experiments

Installation description

The Bedous plant is on the Aspe River (20 km upstream from the Soeix plant), at a point where the mean annual discharge is about 16 m3/s. Mean monthly discharges for March, April, and May are 27, 35, and 43 m^3/s, respectively. The hydroelectric plant consists of a dam that diverts water through a 250-m-long canal, then a 12-km tunnel and a 350-m penstock to the Asasp powerhouse. The powerhouse is equipped with two Francis turbines with a nominal discharge of 14 m^3/s (maximum total turbine discharge 28 m^3/s). The 13.8 MW units have a rated head of 113 m.

The 14-m wide water intake (Figure 3) is situated at the junction between the canal and the tunnel. It is fitted with a trashrack (rectangular bars with a spacing of 3.0 cm) with a submerged cross-section of about 56 m^2. The level at the intake canal fluctuates about 0.30 m during periods of smolt migration.

Flow patterns

Immediately downstream from the conveyance intake, a sharp bend in the intake canal induces a very marked asymmetry in the flow distribution, the highest velocities being at the outer bank of the bend; that is, on the right bank. This nonuniformity in flow is maintained along the entire length of the canal, up to the level of the power intake, where the trashracks are located. The velocities can attain 1.7–1.9 m/s on the right bank. On the left bank, a large recirculation area is spread over about 30 m upstream from the trashrack.

Bypass characteristics

The downstream migration device consists of a surface bypass 1 m wide, fitted with an automatic flap gate (0.70 m in width by 1.50 m in height), controlling the discharge in the bypass and supplying a stilling pool equipped with a trapping system. During the experiment period in the spring of 1995, the discharge into the bypass was regulated successively at 0.4, 0.5, and 0.7 m^3/s, representing 1.7 to 4.3% of the turbine discharge.

Passage tests

A total of 1,900 marked hatchery-reared smolts, divided into 19 batches, were released into the intake canal in the spring of 1995. Three hundred twenty six (17.1%) passed through the bypass. The bypassed percentages varied from 9.2 to 30.6% depending on the batches.

Nineteen individuals were monitored by radiotracking after being released into the intake canal at the same location as the marked fish. Seven of the 19 fish (37%) used the bypass. The current down the intake canal swept the majority of the individuals in the high-velocity flow on the right bank. They arrived in front of the trashracks, generally in the middle of the canal, where transversal flow near the trashrack quickly drew them into the countercurrent on the left bank where they then re-

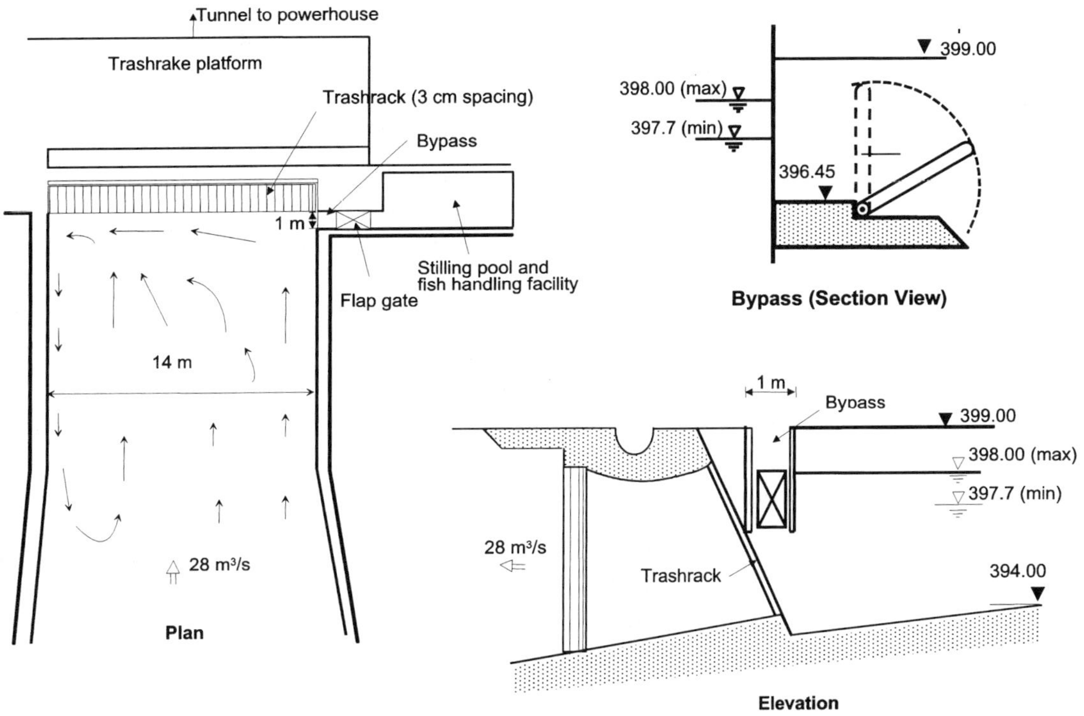

Figure 3. Schematic diagram of Bedous intake (Aspe River) and fish downstream bypass.

mained. They would swim in front of the trashrack while coming back down the main current (in the middle of the right half of the canal). The number of fish that remained near the bypass entrance was not significant, and they spent little time there.

As observed on the Soeix site, hydraulic conditions appeared to be most important factor influencing bypass operation. The bypass location on the bank opposite the area where fish congregated (the countercurrent) was likely the cause of its low efficiency. The only fish that passed through were those that arrived directly along the right bank into the attraction area of the bypass.

When the high mortality rate (more than 50%) of fish passing through the turbines at the Aspe plant is taken into account, the efficiency of the bypass is notably insufficient. Based on the observed hydraulic conditions, the chances of significantly improving the efficiency of the present bypass efficiency seem very slight. One solution would be to build a second bypass on the left bank (which creates a significant construction problem), another in drastically changing the flow conditions in the intake canal by reversing the flow pattern. The second solution would consist of placing a baffle 4 m in length on the right bank. This latter solution was tested on a physical scale model, and appeared to be hydraulically efficient. Tests to verify that this solution significantly improves the efficiency of the bypass are planned for 1998.

St. Cricq Experiments

Installation description

The St. Cricq plant is on the Ossau River, which has a mean annual discharge of 18 m^3/s at that site. Mean monthly discharges for March, April, and May are around 18, 28, and 43 m^3/s, respectively. The St. Cricq hydroelectric installation consists of a gravity dam that supplies the power plant forebay through a 2.5-km tunnel. The power intake supplies, via two penstocks, two Francis turbines with a nominal discharge of 9.5 m^3/s each. The two 6-MW units have a rated head of 62 m.

The forebay is rectangular (Figure 4), and is 11 m wide by 15 m long. Water depth varies from 3 to 4 m depending on the number of turbines in operation. The water intake is fitted with a trashrack (bars 2.5 cm apart) with a submerged cross-sectional area of 40 to 59 m^2 depending on the water level in the forebay.

Flow patterns

Two symmetrical recirculation currents characterize the flow pattern in the forebay. These currents are generated by the high velocity of water (2 m/s) arriving from the tunnel at the center of the basin, and they create two tangential currents at the trashracks, starting from the center of the trashrack and directed towards the right or left banks of the forebay. The tangential velocity at the trashrack is high (0.65–0.80 m/s).

Bypass characteristics

Two adjoining downstream bypasses were constructed on the left bank of the forebay about 0.8 m from the trashrack. They are rectangular-profile, broad-crested weirs (1 m). The upstream corners of the weirs were rounded to reduce flow contraction. The weir crests were set at a 1 m difference in elevation. They operate separately according to the forebay water level, which corresponds to the operation of the plant with one or two turbines. Their characteristics are as follows: 0.8 m wide, and 0.7–0.95 m deep (depending on the changes in water level in the forebay). Discharge from the lower bypass varies from 0.25 to 0.65 m^3/s, corresponding to a range of 1.3% to 3.5% of the turbine discharge. The head difference through the bypass weir is small (5–8 cm); the relatively weak acceleration of current in the bypass that results gives a mean velocity of 0.52 m/s (low level) and 0.83 m/s (high level) at the entrance to the bypass, and 0.55 m/s and 0.91 m/s, respectively, at the exit.

Passage tests

The tests were conducted in 1996 with hatchery-reared salmon smolts (Travade et al. 1997). A total of 1,539 fish divided into 16 batches of 50 to 144 individuals were released into the forebay. Of these, 1,216 fish went through the downstream bypass, for an average efficiency rate of 79%. Depending on the batch, the efficiency of the bypass was between 63 and 100%.

Fifteen smolts were monitored by radiotracking. The efficiency of the bypass for these fish was comparable to that for recaptured, marked fish: 14 smolts (93%) used the bypass. The principal observations obtained by radiotracking included the following:

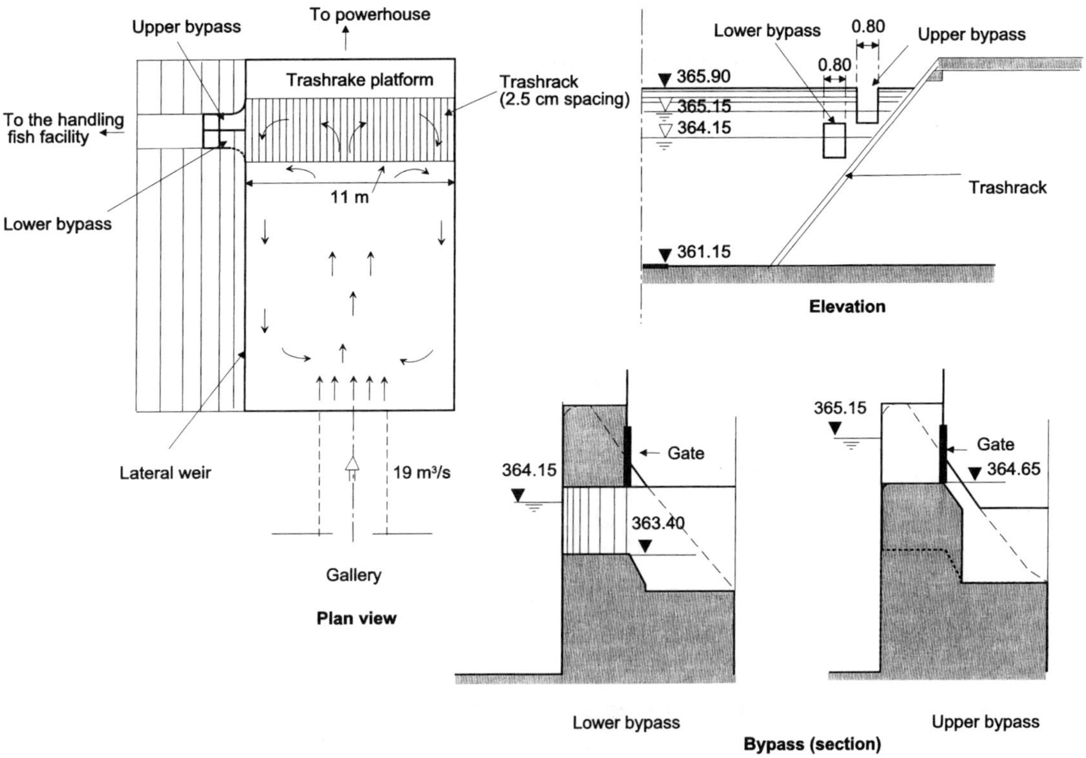

Figure 4. Schematic diagram of St. Cricq intake (Ossau River) and fish downstream bypasses.

- The length of time fish stayed in the forebay in front of the trashracks was very brief: more than half the fish (7 out of 13) went through the bypass in less than 20 s, and only 2 fish stayed in the basin for more than an hour before going through the bypass.
- For the most part, hydraulic conditions in the forebay and more particularly the high velocity component parallel to the trashrack explain the rather high efficiency of the bypass. The fish arriving in front of the trashrack on the left bank were very quickly drawn to the bypass, which they entered within 2–15 s. When the fish arrived from the right bank, they recirculated in the right bank countercurrent until they were drawn to the left bank. Furthermore, some fish moved against the current along the trashrack. The length of time that right bank fish stayed in the forebay was higher (15 s to 80 min) than for those arriving directly along the left bank.
- The small spacing between the bars of the trashracks, as well as the tangential component of the velocity, most probably con tributed to making the trashrack highly repulsive, acting as a louver.
- Lengthy stays of fish in front of the entrance to the bypass were not observed, as had been the case at Soeix and Bedous. It is likely that the significant depth of the bypass as well as the weak velocity gradient (resulting from the weir profile of the bypass and the small head difference through it) suppressed the escape reactions observed at the bypass gate crest.

- Fish passed through the bypass for all rates of bypass discharge (0.25–0.65 m^3/s). There seems to have been no effect, within this range, on the efficiency of the bypass.

Camon Experiments

Installation description

The Camon plant is on the Garonne River. The mean annual discharge is about 60 m^3/s at the intake of the plant. Mean monthly discharges for March, April, and May are around 52, 72, and 92 m^3/s, respectively. The hydroelectric plant consists of a dam that diverts water through a 3.4 km-long canal to the Camon powerhouse. The powerhouse is equipped with three Francis turbines with a nominal discharge of 28 m^3/s (total turbine discharge 85 m^3/s). The 3-MW units have a rated head of 20 m.

The water intake (Figure 5), 30 m wide, is fitted with a trashrack (rectangular bars with a spacing of 4.5 cm) with a section of 190 m^2. The changes in level in the intake canal are approximately 1 m during downstream smolt migration. On the left bank, along a width of 3 m, the bar-spacing of the trashracks is wider (17 cm). This zone supplies a conduit which in turn supplies the intake of the Valentine plant located in the tailrace 3 km downstream in case of a sudden shutdown of the Camon plant.

Bypass characteristics

The downstream bypass device built in 1996 is above this conduit. It consists of a sluice gate 2.3 m in wide supplying a stilling pool. The trapping device is downstream from the stilling pool. The crest level automatically adjusts to the intake canal water level fluctuations. During the test period, the discharge through the gate was between 2.5 and 4 m^3/s.

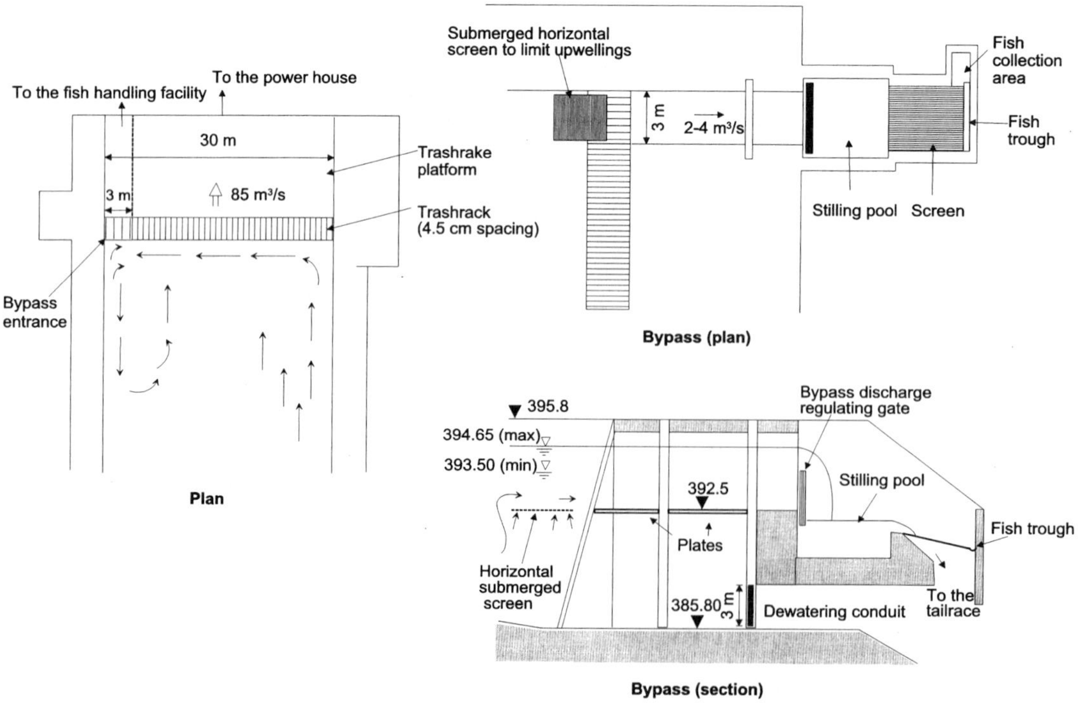

Figure 5. Schematic diagram of Camon intake (Garonne River) and fish downstream bypass.

Flow patterns

Mean velocity in the intake canal can vary considerably according to turbine discharge (0.5 m/s for maximum turbine discharge). Flow patterns in the intake canal are not symmetrical. Maximum velocities occur on the right bank (up to 1.20 m/s); velocities are less than 0.6 m/s at a distance of 5 m from the left bank. We noted a tangential flow along the trashrack (maximum velocity of 0.75 m/s), with the flow pattern extending from the right bank to the left bank. However, this transverse current is suddenly interrupted at a distance of 3–5 m from the left bank by an unstable upwelling. On the left bank, a recirculation area exists that extends to about 40 m upstream from the trashrack.

Passage tests

A first series of tests was carried out in 1996 with hatchery-reared smolts (Carry et al. 1996). A total of 2,025 smolts, divided into 11 batches, were released into the intake canal about 600 m upstream from the plant. Only nine batches were analyzed, due to operational incidents that occurred at the bypass gate. With 1,625 fish accounted for, 550 (34%) passed through the bypass. Percentages varied from 6.5 to 76% depending on the batches.

Eleven smolts were monitored by radiotracking after being released into the intake canal and only three fish (27%) passed through the bypass.

The length of time necessary for fish to swim between the release point to the trashrack was very small, from 5 to 11 min. Fish arrived in the observation area mostly from the right bank. The transverse current drew them towards the middle of the canal, then generally to the left bank. Smolts then stayed in front of the bypass in an area 5–6 m wide. The length of time they stayed in front of the trashrack was highly variable (i.e., from 4 min to more than 3 h). All of the radio-tagged fish returned at least three times to the bypass entrance and spent most of their time in this area, indicating proper location of the bypass. However, the repeated return of smolts to the bypass entrance before they passed through the trashracks indicated that the device was not sufficiently attractive. The 4.5 cm bar-spacing of the trashrack would appear to be dissuasive, but in fact this dissuasive effect was probably increased by the tangential flow along the trashrack.

A second experiment was carried out in 1997 (Carry et al. 1997), in which an attempt was made to improve the hydraulic conditions at the bypass entrance by installing horizontal, submerged metal plates (Figure 5) between the bypass entrance and the bypass gate to increase the horizontal velocity in this area. Also, a submerged horizontal screen was installed just a few meters upstream from the trashracks (size: 3 m × 3 m and porosity close to 20%) to reduce the upwelling at the bypass entrance.

The tests were made with hatchery-reared smolts: 1,528 smolts divided into 16 batches were released into the canal intake about 600 m upstream from the power plant. Of these, 1,119 (73%) passed through the bypass. The percentages varied from 57% to 84% depending on the batches. This increase in the efficiency of the downstream migration device was partly due to the improvement of flow pattern conditions just upstream from the bypass. However, it was difficult to attribute the improvement only to these changes: the turbine discharge was much lower in 1997 (30–65 m^3/s) than in 1996 (70–85 m^3/s), and fewer turbines were operating, resulting in a decrease in intake canal velocities: 0.75–0.85 m/s on the right bank instead of 1.0–1.20 m/s.

Lighting effect

The effect of lighting on the passage of smolts was dramatic: smolts passed in significantly higher numbers during the 2–5 min after the lamp was switched off. Radiotracking also showed the attractive effect of the lamp at the bypass entrance, with radio-tagged fish staying near the lighted area while the lamp was turned on (Figure 6).

Discussion

Experiments conducted over a six-year period at Soeix, Bedous, Camon, and St. Cricq enabled successive testing of seven downstream bypass configurations designed to provide safe passage for salmonid juveniles at turbine intakes. The devices tested were surface bypasses installed on the riverbank near the trashracks of the power intake or to the side of the

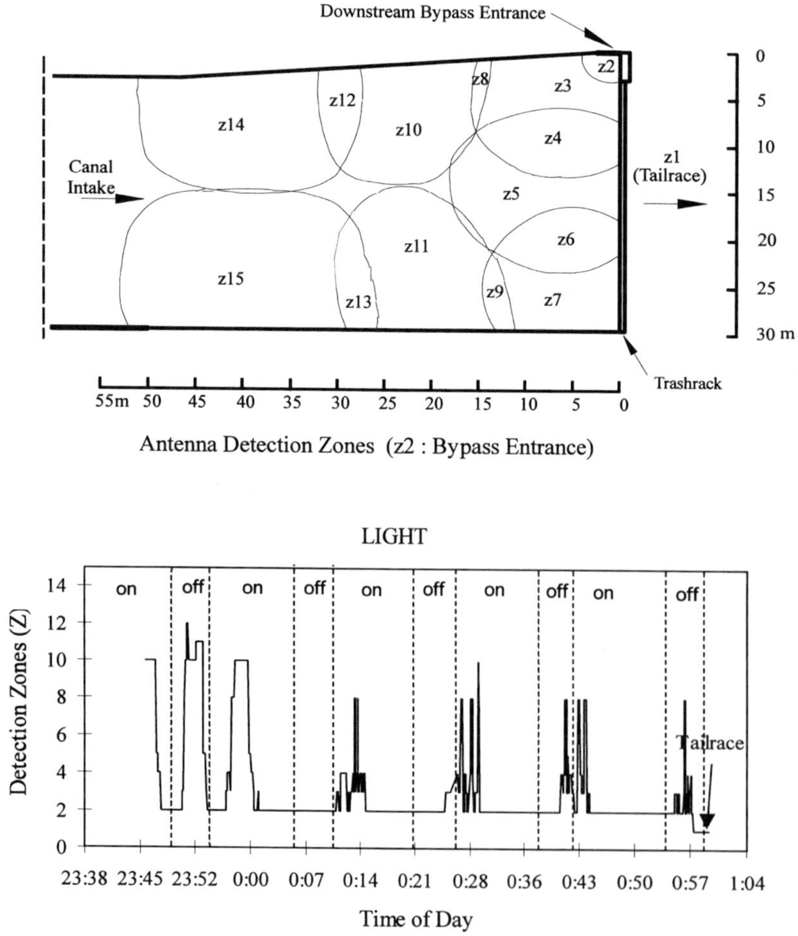

Figure 6. Example of fish displacement in reaction to intermittent lighting of the downstream bypass at Camon intake.

trashracks themselves. The use of two complementary techniques, mark–recapture and fish radiotracking, linked with a description of hydraulic characteristics (in front of the trashracks and near the bypass entrances) made it possible to assess bypass efficiency and to highlight those factors that have a bearing on fish behavior.

We found that the mean efficiency of the downstream migration bypasses varied greatly depending on the sites and the characteristics of the devices: 17% at Bedous and close to 80% at St. Cricq. This variability appears to be related to several factors that will be explained in detail.

Trashrack characteristics and hydraulic conditions

It was clear that the trashracks (bar-spacing from 2.5 to 4.5 cm), had a repulsive effect on smolts. In effect, the majority of fish slowed down their migration (from a few minutes to a few hours), when they encountered trashracks. However, a certain number of fish (e.g., 10% at Soeix) did not pause and passed directly through. It seems obvious that the dissuasive aspect of the trashracks will decrease with the increase of the spacing between the bars.

The characteristics of the trashrack (bar-spacing) and velocity pattern in front of the trashrack, in relation to the size and the swimming ability of the fish, influenced the potential for dwelling in front of the trashracks, and, consequently, the probability that fish will find the bypass entrance. The mean length of the marked fish that passed through the bypass was generally significantly greater than that of the released fish. This difference indicates that the repulsiveness of the trashrack is dependent on the ratio between fish size and bar spacing.

The existence of a well-marked component of velocity parallel to the trashrack (guiding the fish by a "louver effect") appeared to significantly increase its repulsiveness. The high efficiency of the bypass at St. Cricq was due not only to the small bar-spacing of the trashracks (2.5 cm) but also to the fact that the high tangential velocity (0.6 m to 0.8 m) very quickly guided the fish towards the bank where the bypass was located. At Camon, the tangential velocity guiding the fish along the trashracks was effective in spite of the greater bar-spacing of the trashracks (4.5 cm).

It is obvious that there is a limit to the length of trashracks, above which several bypasses are necessary. The satisfactory results from Camon in 1997 would tend to show that one bypass may be enough for a 30-m trashrack. However approach flow conditions must also be taken into account. The flow patterns were probably very favorable at Camon, with a tangential current at the trashracks guiding fish towards the bypass.

Location of the bypass

If we take into account the behavior of fish staying very close to the trashracks, it appears that the bypass entrance should be as close as possible to the trashracks (less than 2 m) if not built into the trashrack itself at one of its edges. The location of the bypass depends strictly on which side of the channel the fish are concentrated, which itself depends on hydraulic conditions upstream from the trashracks (tangential currents, recirculation areas, upwellings, etc.). The Bedous tests were instructive in this respect. The position of the bypass opposite the bank with the countercurrent was determined to be the main factor responsible for the low efficiency of the installation. Radiotracking showed that fish congregated and essentially stayed in the countercurrent on the left bank, while very few of them gathered in the right bank bypass area.

Bypass discharge

Although discharge was not tested systematically, we believe that discharge in the bypass should be related to the turbine discharge. As for upstream fishpasses, the minimum required discharge appears to be a few percent of the turbine discharge. The importance of bypass discharge appears subordinate to bypass location and hydraulics conditions. The good results obtained in 1994 at Soeix were for a discharge that was 2–10% of the turbine discharge (median 5%). In this respect, it appears very important to take fluctuations in upstream water level into account when determining the level of the bypass crest, so as to maintain a constantly adequate discharge. These criteria for discharge remain dependent on other parameters influencing the bypass efficiency (location of the bypass, flow patterns, trashrack characteristics); that is to say, if the other parameters are unfavorable, it may be necessary to increase discharge in the bypass. Thus, at St. Cricq, where very favorable hydraulic conditions led the fish very quickly to the bypass entrance, the discharge of 1.3–3.5% appeared to be sufficient to ensure bypass efficiency. At Camon, percentages of bypass discharge in relation to turbine discharge ranged from 4 to 13% in 1997 (73% mean efficiency).

Hydraulic conditions at the bypass entrance

The hydraulic conditions in the immediate vicinity of the bypass entrance appear to be essential to ensure adequate fish passage. These conditions must include undisturbed surface flow (neither flow separation from the boundaries nor turbulence) and be perceptible from as great a distance as possible from the bypass, with a gradual acceleration up to the bypass entrance.

Accelerated flows must occur both on a horizontal and a vertical plane, hence the importance of the geometry of the bypass entrance (no sharp angles, use of surface flap gates or broad-crested weirs rather than flat vertical gates). When possible, a broad crested weir should be used to create a gradual acceleration in the flow pattern, and to ensure a small head difference (lower than 10 cm) in the bypass with the aid of a regulating pool downstream from the bypass. This seems to be why St. Cricq was the most satisfactory of the bypasses tested, with fish showing no hesitation at all before passing.

Effect of lighting in attracting fish to the bypass

Visual observation and radiotracking showed that smolts are attracted into the lighted areas. However, we found no significant increase in bypass efficiency for intermittent illuminated conditions. We believe that hydraulic conditions are indeed the most important factor and that light does not noticeably affect bypass efficiency, whether the hydraulic conditions are very favorable or whether they are very limiting. It would be worthwhile to test this hypothesis in future sites where hydraulic conditions are not so significant.

Conclusions

Experiments suggest that surface bypasses, associated with conventional trashracks, can be effective in passing juvenile salmonids around small-scale hydroelectric plants; passage efficiencies as high as 60–80% have been observed on certain sites.

Several factors have been identified as important to the success of such bypasses:

- trashrack of limited spacing (maximum 3–4 cm);
- a lack of uniformity in approach flow conditions, generated most of the time by lack of symmetry in the geometry of the intake canal, creating a louver effect with the bars of the trashrack and guiding fish towards the bypass;
- a suitable bypass location (close to the trashrack plane, where fish tend to congregate) and discharge (several percent of the turbine discharge); and
- suitable hydraulic conditions in the zone of influence and at the bypass entrance (e.g., absence of upwelling, limited acceleration, and turbulence).

At very small hydroelectric plants, it seems possible to modify the hydraulic pattern in the intake canal (by installing deflectors) to make the use of a surface bypass more efficient. Like other behavioral barriers, performance is site-specific and it is difficult to predict the efficiency of such a device; only on-site evaluation can give the response. The principle of such surface bypasses, in which the bars of the trashrack do not physically prevent fish from entering the intake, is purely behavioral; as a result, efficiency will always be limited and their use on existing projects will be restricted only to those where the bar spacing and flow pattern in the canal are favorable.

In the French context, where in most cases it does not seem necessary to provide 100% downstream protection, surface bypasses associated with existing trashracks may be an acceptable mitigation technology. However, where it is necessary to guarantee a highly efficient downstream passage (more than 80%, for example, in the case of drastic turbine mortality), more standard and costly solutions should be considered, such as physical barriers or even transportation if there are numerous projects on the same river.

Acknowledgments

These studies were carried out in collaboration with the CSP (National Council of Inland Fisheries), CEMAGREF (French Institute of Agricultural and Environmental Engineering), and EDF (Electricité de France). Funding was provided by the Regional Administration, CSP, and EDF. We particularly thank J. M. Bach, D. Barracou, L. Carry, M. Chanseau, O. Croze, N. de Faveri, E. Galiay, R. Galland, C. Gouyou, D. Ingendahl, and D. Pujo for their contribution to the different field studies.

References

Armstrong, J. D., and C. E. Rawlings. 1993. The effect of intragastric transmitters on feeding behaviour of Atlantic salmon, *Salmo salar*, parr during autumn. Journal of Fish Biology 43:646–648.

Bomassi P., and F. Travade. 1985. Projet de réimplantation du saumon dans la partie supérieure de l'Allier: expériences sur les possibilités de dévalaison des saumoneaux au barrage hydroélectrique de Poutès en 1983 et 1984. Colloque pour la restauration des rivières à saumon, Bergerac, 28 mai-1er juin 1985 (in French).

Carry, L., M. Chanseau, O. Croze, E. Galiay, and M. Larinier. 1996. Expérimentation d'un dispositif de dévalaison pour les juvéniles de saumon atlantique (Usine hydroélectrique de Camon - Garonne année 1996). Rapport GHAAPPE 96.07 (in French).

Carry, L., E. Galiay, M. Larinier, and G. Oules. 1997. Expérimentation d'un dispositif de dévalaison pour les juvéniles de saumon atlantique (Usine hydroélectrique de Camon - Garonne année 1997). Rapport GHAAPPE 97.03 (in French).

Clay, C. H. 1995. Design of fishways and other fish facilities. Lewis Publishers, Boca Raton, Florida.

EPRI (Electric Power Research Institute). 1986. Assessment of downstream migrant fish protection technology for hydroelectric application. EPRI RP 2694-1, Palo Alto, California.

EPRI (Electric Power Research Institute). 1994. Research update of fish protection technologies for water intakes. EPRI TR-104122, Palo Alto, California.

Larinier, M., and S. Boyer-Bernard. 1991a. Dévalaison des smolts et efficacité d'un exutoire de dévalaison à l'usine hydroélectrique d'Halsou sur la Nive. Bulletin Français de Pêche et Pisciculture 321:72–92 (in French).

Larinier, M., and S. Boyer-Bernard. 1991b. La dévalaison des smolts de saumon atlantique de Poutès sur l'Allier (43): utilisation des lampes à vapeur de mercure en vue d'optimiser l'efficacité de l'exutoire de dévalaison. Bulletin Français de Pêche et Pisciculture 323:129–148 (in French).

Larinier, M., and F. Travade. 1996. Smolt behavior and downstream fish bypass efficiency at small hydroelectric plants in France. Pages 891–902 *in* M. Leclerc, and four coeditors. IARH (International Association for Hydraulic Research) 2nd symposium on habitats hydraulics. Ecohydraulics 2000. Institut Nationale de Recherche Scientifique, Quebec.

Moser, M. L., A. F. Olson, and T. P. Quinn. 1990. Effects on dummy transmitters on juvenile coho salmon. Pages 353–356 *in* N. C. Parker, and five coeditors. Fish-marking techniques. American Fisheries Society, Symposium 7, Bethesda, Maryland.

Ruggles, C. P. 1980. A review of the downstream migration of Atlantic salmon. Canadian Technical Report of Fisheries and Aquatic Sciences 952:1–39.

Ruggles, C. P. 1992. What's new in downstream fish passage? Pages 402–416 *in* D. Mills, editor. Salmon in the sea and new enhancement strategies. Fourth international Atlantic salmon symposium. Fishing News Books, Oxford, UK.

Taft, N. 1988. Evaluations of fish protection systems for use at hydroelectric plants. HydroReview VII(2):54–62.

Travade, F., and six coauthors. 1989. Use of radiotracking in France for recent studies concerning the EDF fishway program. Hydroécologie Appliquée 1(2):33–51 (in French).

Travade, F., N. De Faveri, C. Gouyou, D. Barracou, and R. Galland. 1997. Expérimentation d'un dispositif de dévalaison pour les juvéniles de saumon atlantique (année 1996). Usine hydroélectrique de St Cricq (Gave d'Ossau - 64). Rapport EDF-DER (in French).

Brian N. Hanson

Effectiveness of Two Surface Bypass Facilities on the Connecticut River to Pass Emigrating Atlantic Salmon Smolts

3

New England Power Company constructed downstream passage facilities at the Vernon and Bellows Falls Hydroelectric Stations on the Connecticut River as part of a 1990 agreement with the Connecticut River Atlantic Salmon Commission to provide safe, timely downstream passage for emigrating Atlantic salmon *Salmo salar* smolts. These passageways are very different in terms of hydraulic and structural characteristics and they offer a striking contrast in the success of their respective surface bypass systems.

Vernon Station is run-of-the-river with nominal generating capacity of 27 MW. Total discharge is up to 325.64 m^3/s at normal operating head of 10.4 m. A log/ice boom forms an inner forebay and diverts floating debris and ice to a sluice at the eastern end of the powerhouse. Initially, this boom and sluice was the only downstream fish passage facility, other than spill gates. Over the course of six years, 1990–1996, studies of radio-tagged emigrating salmon smolts were conducted at the Station. Resultant data were used to facilitate modifications and additions to the Station to provide safe, expeditious downstream passage. A 25-m-long, surface-fed fishpipe bypass was installed through the powerhouse in late 1990 and studied for two years. A louver system to guide smolts to the fishpipe and an alternate surface bypass was constructed in 1993 and studied in 1994, 1995, and 1996. Bypass efficiency was improved and the agencies approved Vernon's downstream facilities for current Station configuration.

Bellows Falls Hydroelectric Station is 51.5 km upstream of Vernon. A 196-m-long by 13-m-high dam diverts the river down a 470-m-long canal to the powerhouse. Total turbine discharge is approximately 297.3 m^3/s. An existing wooden log/ice boom was replaced in 1994 with a fixed concrete diversion boom that extends at a 45° angle from one side of the power canal to the entrance of a sluice. The boom is 62.8 m long and extends 4.6 m underwater at normal impoundment elevation. Water passes through the sluice over a regulated skimmer gate through a 76-m straight-walled concrete tunnel. Elevation drops approximately 9 m along the sluice and at the end, water falls from a height of 9 m into the tailrace. This diversion boom system successfully bypassed 94.4% of all radio-tagged salmon. After only one year of study, in 1995, (at least three years were initially required), the agencies approved the Bellows Falls downstream facility.

Introduction

Passage of emigrating anadromous fishes at hydroelectric facilities on the Connecticut River has long been a concern of resource agencies and power utilities. On 26 July 1990, New England Power Company (NEP), and the Connecticut River Atlantic Salmon Commission (CRASC) and its member agencies, signed a Memorandum of Agreement to establish a schedule for the resolution of the downstream passage issues at selected dams. Two of the facilities, the Vernon and Bellows Falls Hydroelectric Projects, were studied using radio-tagged hatchery and stream-raised (wild) emigrating Atlantic salmon *Salmo salar* smolts.

The utilization by salmon of existing passage routes (tainter gates and log/ice sluice) at Vernon Station was monitored for six years beginning in 1990. Subsequent studies were conducted to determine the effectiveness of new bypass facilities installed from 1991 through 1994. These included a surface-fed fishpipe, installed in 1991, and a louver system and alternate western forebay bypass, installed in 1993. The fishpipe, which extended through the middle of the powerhouse to allow passage from the Station's forebay to the tailrace, was studied during 1991 and 1992. Results of these studies indicated that emigrating radio-tagged smolts generally did not utilize the downstream routes (log/ice sluice and fishpipe) in proportions desired. Highly variable water currents within the forebay and higher intake flows in the western forebay area tended to complicate the attraction of the downstream passage facilities and tended to draw smolts to the western intake areas where they prolonged for extensive periods. Methods implemented (low-frequency ensonification) to alleviate these problems and to guide smolts to available passage routes were largely unsuccessful. In response, NEP constructed two new structures in 1993 in an attempt to facilitate downstream passage. A louver system was installed within the forebay to guide emigrating fish to the entrance of the fishpipe, and a fishtube, with a surface fed entrance, was installed in the western portion of the forebay to offer a downstream route for fish congregating in that area. To examine the effectiveness of these facilities, the log/ice sluice was closed and attempts were made to monitor radio-tagged smolts during periods of no spill. The louver guidance/fishpipe facility and the western fishtube bypass were studied from 1994 through 1996.

At Bellows Falls Hydroelectric Station, emigrating salmon smolts are afforded two passage routes during periods of no spill at the dam: a log/ice sluice and turbines. An earlier study indicated that the log/ice sluice was a safe passage route for emigrating Atlantic salmon smolts (RMC 1991). Emigration of radio-tagged salmon was studied informally during 1992 and 1993. Data suggested that the existing floating boom in the forebay, which diverted ice and debris to the concrete sluice tunnel, was not an effective passage route. These results prompted investigation of measures to divert emigrating smolts to the entrance of the sluice. After extensive modeling studies at Alden Research Laboratory (Worcester, Massachusetts), a new structure was designed to replace the existing log boom immediately upstream of the powerhouse intakes. Construction of the new fish diversion boom commenced in late 1993 and was completed in late spring of 1994. The structure was intended to facilitate passage of emigrating fish through the log/ice sluice, and exclude fish from the area of the turbine intakes. As an additional measure, a secondary entrance to the sluice was constructed on the eastern wall, just upstream of Unit 3 intake, for emigrants that might pass under the boom. Extensive study of the effectiveness of this bypass system was conducted in 1995.

Site Descriptions

Vernon Dam and Bellows Falls Dam are on the Connecticut River at river km 228.5 in the towns of Vernon, Vermont and Hinsdale, New Hampshire and at river km 280 in the towns of Rockingham, Vermont and North Walpole, New Hampshire, respectively (Figure 1).

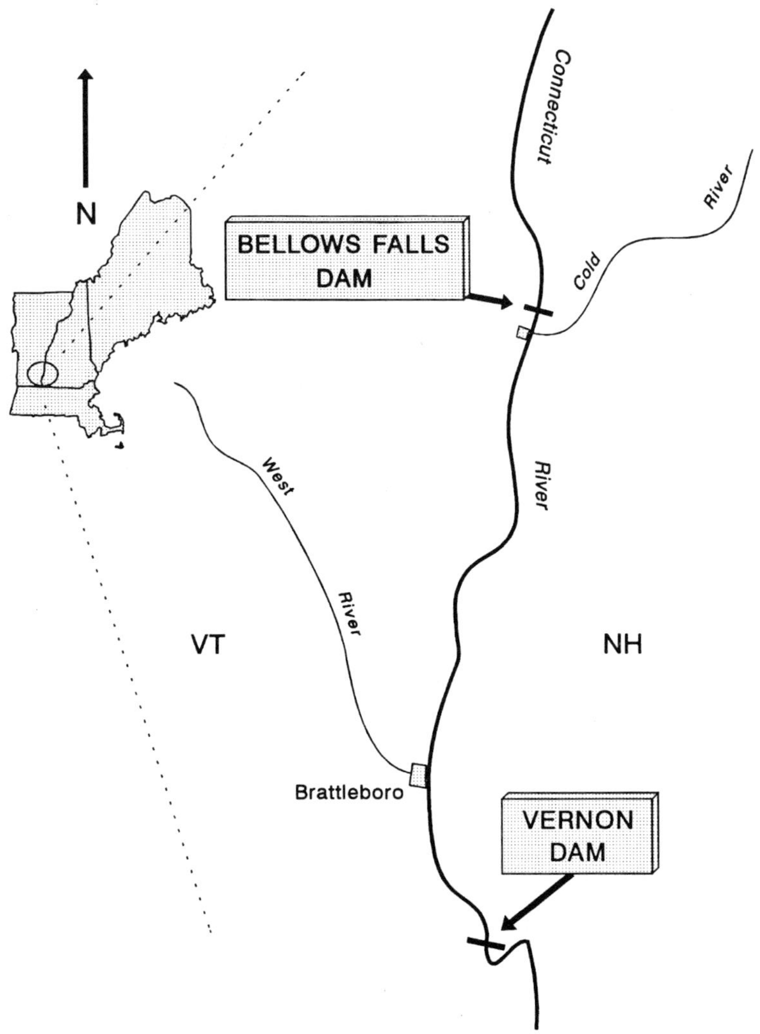

Figure 1. Schematic map showing locations of Vernon and Bellows Falls Dams.

Vernon project

Vernon Dam is a concrete gravity structure and was constructed in 1909. The station is a run-of-the-river facility with a nominal generating capacity of 27 megawatt (MW). Ten Francis generating turbines are housed in the powerhouse:

- four turbines (Units 1–4) have single runners with a maximum discharge of 36.2 m^3/s each,
- four turbines (Units 5–8) have triple runners with discharge rates of 27.7 m^3/s each, and
- two turbines (Units 9 and 10) are larger single runner Francis types with a maximum discharge of 52 m^3/s each.

Units 5 and 8 became inoperable and were retired during the study years. Normal operating head is 10.4 m. The turbine intake trashrack bars are spaced 5 cm apart for Units 1 through 8 and 10 cm apart for Units 9 and 10.

The dam creates an impoundment with a surface area of 10,320 hectares (ha) at a normal elevation of 67 m above mean sea level. Total length of the dam is 291.3 m; the eastern 182.9 m contains 6 tainter gates. Also in this area are eight 2×2.7 m underwater flood gates used to regulate smaller amounts of surplus water. Each flood gate can pass approximately 57 m^3/s into the tailwater of the dam. A log/ice boom, anchored to 5 concrete piers, forms an inner forebay area and diverts trash and ice to a 4-m-wide, 27.4-m-long concrete sluice at the eastern end of the powerhouse (Figure 2). A skimmer gate (gate opens downward underwater to allow overflow) regulates flow down the sluiceway. Flow varies from 1.2 to 78.4 m^3/s at heads from 0.3 to 4.9 m, respectively. The gate is normally set at 1.1 m head (approximately 8 m^3/s) for fish passage. Initially, this sluice (with the exception of spill gates when spilling) was the only downstream passage route for emigrating salmon smolts, other than the turbines. A fishpipe, installed in 1991, is surface fed and allows downstream passage of fish from the inner forebay to the tailrace. The fishpipe measures approximately 25 m in length, is 1.2 m high throughout its length and measures 2.3 m wide at the upstream end tapering to approximately 0.8 m at the downstream end. Flow through the pipe when open is 9.9 m^3/s.

A 47.5-m-long louver system, installed in 1993, in the inner forebay extends from a log boom pier to the entrance of the fishpipe (Figure 3). Eleven 3.7-m-wide by 3-m-high removable louver panels are housed in the system. Each individual stainless steel vertical louver measures 5-cm-wide, 1-cm-thick and 3-m-high and is angled 60° to the axis of the louver system structure; louvers are spaced 7.6 cm apart. At normal impoundment elevation, the louver panels extend 4.6 m below the water surface. The second passage facility utilized (with modifications) an existing conduit, which formerly supplied attraction flow for the fish ladder collection gallery, to attract and pass fish. The surface-fed entrance of the conduit (termed the West Fishtube) was constructed in the western forebay wall of the Station, approximately 7.6 m upstream from the Unit 10 trashrack (Figure 3). The entrance measures 2.4-m-high by approximately 1-m-wide, and at normal impoundment elevation, the bottom sill is approximately 1.8 m underwater. Approximately 1.2 m from the lower sill of the entrance, the passage tapers from 1.2 m in diameter to 0.6 m in diameter, and extends through the powerhouse to discharge in the tailrace. Flow through the fishtube is approximately 1.4 m^3/s.

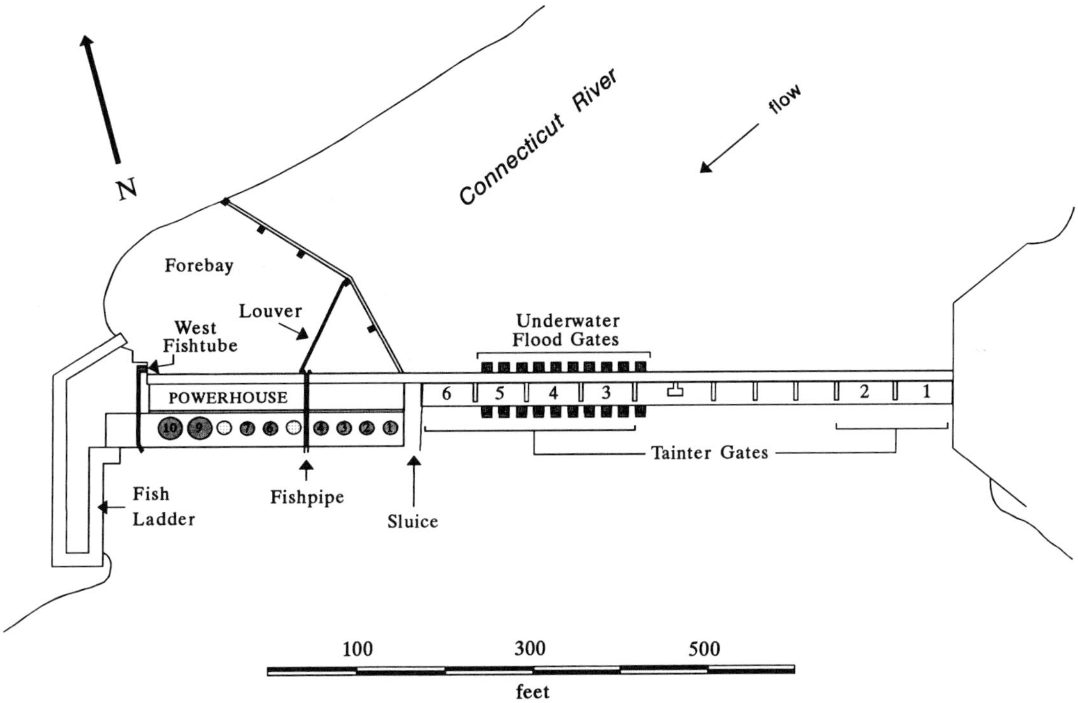

Figure 2. Schematic of Vernon Dam showing passage routes.

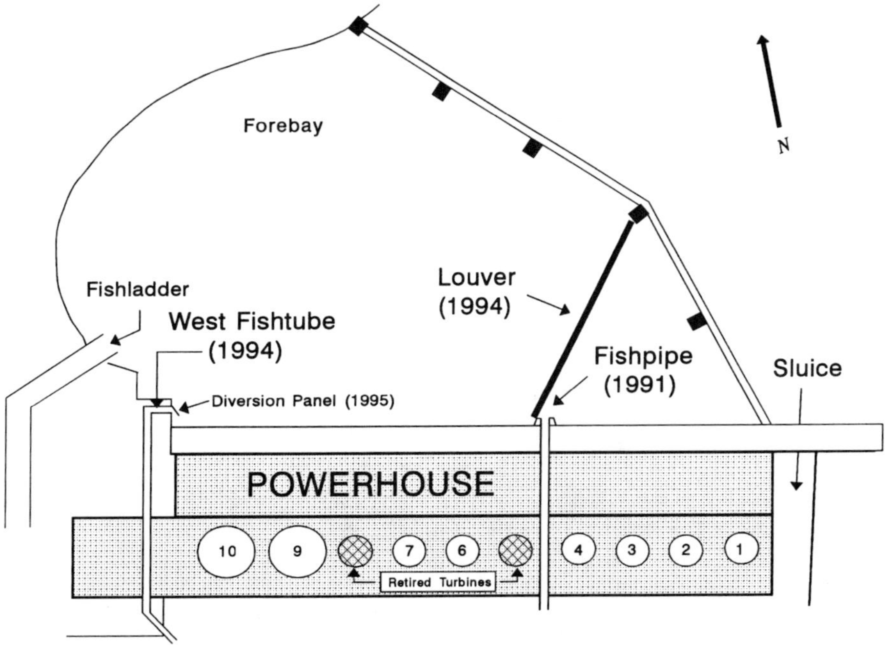

Figure 3. Schematic of Vernon Dam forebay showing passage systems and first year of study.

Bellows Falls project

At Bellows Falls Project, a 196-m-long by 13-m-high concrete gravity dam diverts the majority of the river down a 470-m-long canal to the powerhouse (Figure 4). The dam was reconstructed in 1928 and contains two 35-m-long by 4-m-diameter roller gates capable of discharging up to 2,973 m^3/s. The dam forms an impoundment with a surface area of approximately 11,330 ha at normal impoundment elevation of 88.7 m above mean sea level.

In the powerhouse, three 16 MW Francis turbines generate power at a normal head of 19 m. Each can discharge up to approximately 99 m^3/s. The powerhouse is a steel and masonry structure approximately 56.7 m long. A floating wooden log/ice boom was replaced in 1994 with a fixed cement diversion boom that extends from the fish ladder exit to the log/ice sluice entrance (Figure 5). The boom is approximately 63-m long and extends 4.6 m underwater at normal impoundment elevation. Four 1.5 m by 4.6 m concrete piers support the structure.

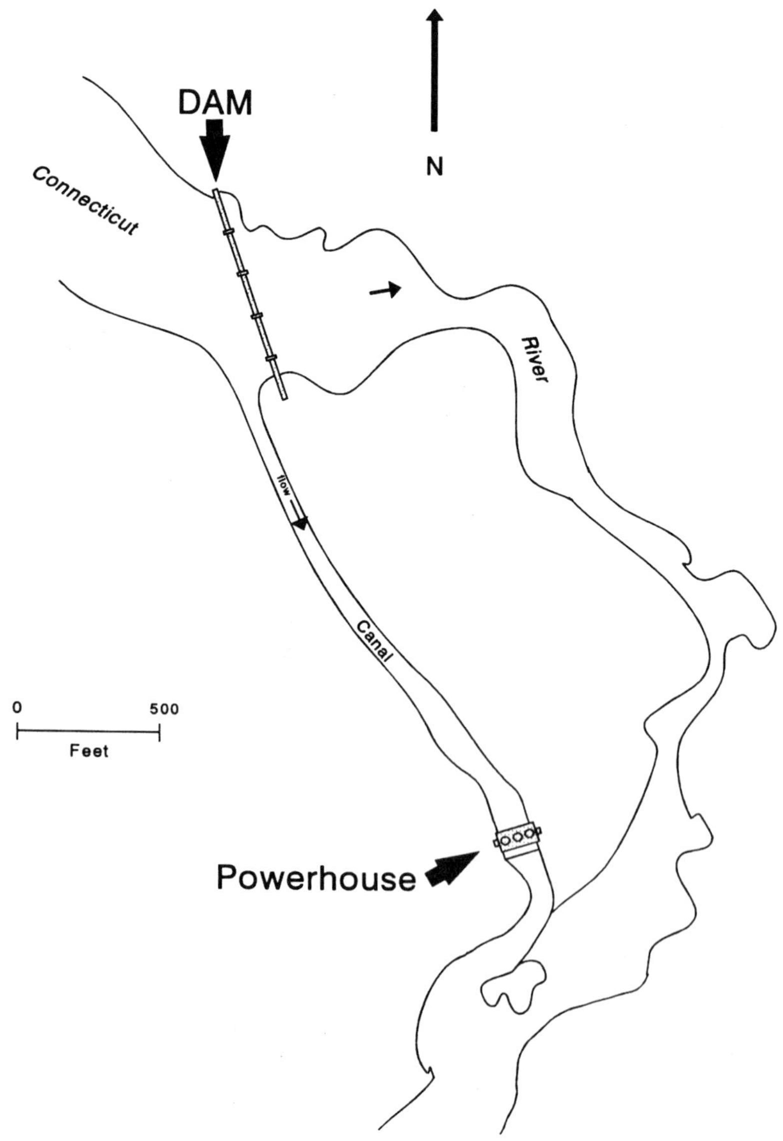

Figure 4. Schematic of Bellows Falls Project.

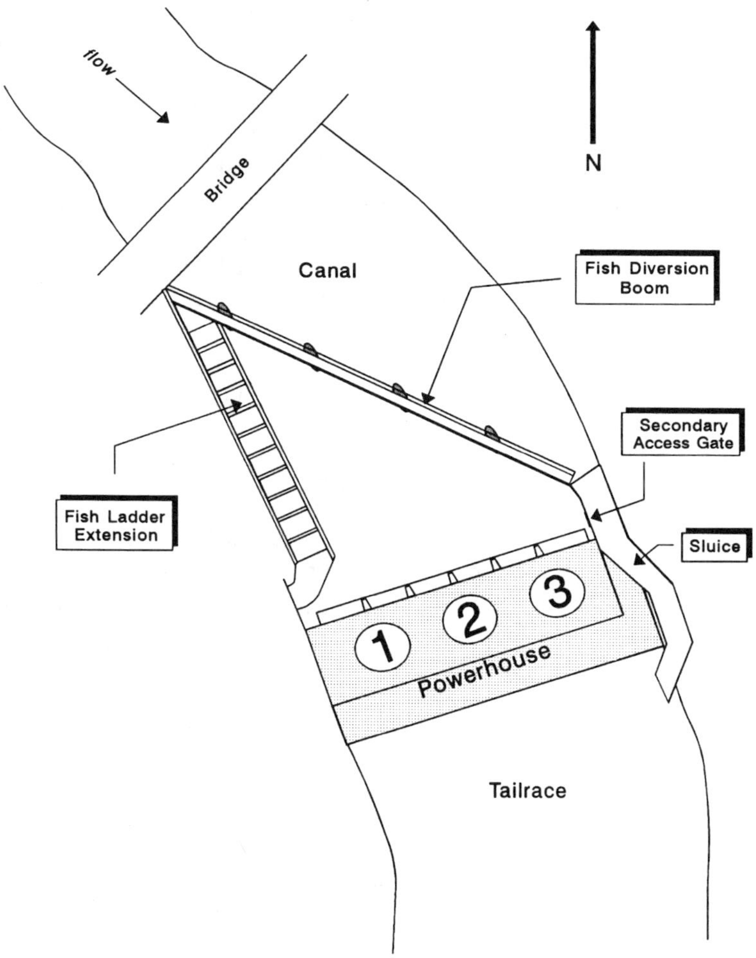

Figure 5. Schematic of Bellows Falls Project forebay showing the bypass system.

Water passing through the log/ice sluice is regulated by a skimmer gate (1.0–28.3 m^3/s at 0.3 and 2.7 m gate openings, respectively). The gate is set at 0.9 m (5.7 m^3/s) for fish passage. The sluice is a 76.2 m straight-walled concrete tunnel measuring approximately 2.7 × 2.4 m at the entrance and tapering to approximately 1.5 × 2.1 m at the discharge. Elevation drops approximately 9 m along the sluice and at the discharge, water falls from a height of 9 m into the tailrace.

Methods

During the six years of study at Vernon Station, the movement and general behavior of radio-tagged emigrating salmon smolts were monitored with strategically placed radio receiving stations.

Smolts were tagged with small radio transmitters procured from Lotek Engineering Inc., Newmarket, Ontario, Canada. Each tag was approximately 1.8 cm in length, 0.8–1.0 cm in diameter, and weighed less than 2 g in water. Signals were transmitted via a short length whip antenna. During the earlier studies at Vernon Project, tags were pulsed at rates of 90, 102, 118, 139, and 170 beats per minute (bpm). Later studies utilized coded pulse tags. Frequencies of transmitters were within the 149–150 megahertz (MHz) bandwidth.

Smolts tagged for the study included age-2 smolts from Nashua Hatchery, age-1 smolts from White River Hatchery and smolts collected from the sluice sampling collection facility (wild smolts) at Northeast Utilities Service Company's Cabot Hydroelectric Station in Turners Falls, Massachusetts. Smolts were held in circular 1.8-m-diameter by 0.9-m deep fiberglass holding tanks installed at Vernon Station. Smolts to be tagged were netted three to five fish at a time from holding tanks and placed in a bath of tricane methanesulfonate (50 mg/L). A radio transmitter was inserted through the mouth and esophagus and deposited in the stomach of each fish. Tagged smolts were measured to the nearest 1 mm fork length (not weighed) and transferred to a transport tank for recovery. All smolts tagged were large enough to accept the transmitter with no adverse effects. Tagged smolts for the Vernon studies were held overnight and released at various locations in the river. In 1990, smolts were released just below Bellows Falls Station (51.5 km upstream of Vernon). During 1991, the release site was moved to an area no more than 0.8 km upstream of the Project to increase radio-tagged smolt detections. In addition, smolts (N = 20) were twice released directly into the forebay in an attempt to increase exposure to the fishpipe. In 1992, the first group of smolts (N = 19) were released just below Bellows Falls Station and the last three groups were released from an area 20 km upstream of Vernon Project, to expedite arrival at the dam. Smolts were released approximately 1 km upstream of Vernon Station during 1994 and during the final two years' studies, all were released 6.6 km upstream of Vernon.

Smolts to be released for the Bellows Falls study were transported to a holding pool at the dam just upstream of the powerhouse and held overnight. The following day they were inspected for fitness and those that appeared normal were released into the river at the dam.

At Vernon Project, Lotek SRX 400 receivers were used to detect the presence of radio-tagged smolts along the spillway, in the log/ice sluice, in the fishpipe, near the intakes, along the louver, in the west fishtube, and in the tailrace. During 1990 through 1994, up to three receivers were deployed to monitor each of the sluice, fishpipe, and west fishtube passage routes. During 1995 and 1996, coded tags were used and DSP 500 units were deployed to monitor the fishpipe and fishtube.

A sonic deterrent system was added to the 1992 study in an attempt to prevent smolts from entering the inner forebay. A low frequency transducer was mounted on the eastern portion of the log boom to deter those smolts migrating westerly along the dam from entering the inner forebay and to direct them to the log sluice for passage.

A total of four monitoring stations was deployed at Bellows Falls Project to detect presence of tagged smolts along the guidance boom, in the sluice tunnel, near the intakes and secondary sluice entrance, and the tailrace. A DSP 500 was utilized to monitor the sluice.

Results and Discussion

Vernon project

During the first three years of study, 1990–1992, 247 radio-tagged salmon smolts were detected at the Project (Table 1). In 1990, only 42 of 122 smolts released were detected. This may be due to many factors, including tag regurgitation, tag malfunction, increasing water temperature, smolt condition, and possible predation. The river characteristics just upstream of the project are such that most smolts were expected to approach from the east or mid-river and traverse the dam westerly. High river flows occurred during the study and 42.9% of the detected smolts passed through spill gates. Those that did not, should have passed the sluice before entering the forebay. Based on radiotelemetry data, 90% of the smolts detected did approach from east- and mid-river; of the smolts that did not pass spill gates, however, 37.5% passed via the log sluice, 54.2% passed through turbines and 8.3% passed by undetermined routes (Table 1). The greatest flow volume in the inner forebay was generated by Units 9 and 10, located in the westernmost portion of the intake area. Many smolts congregated near Units 9 and 10 for up to 4 d before passing through the turbines. Mean residency time, including those that passed other routes, was 8 h 9 min.

Table 1. Summary of passage of radio-tagged salmon smolts results at Vernon Hydroelectric Station during all years of study.

	1990		1991		1992		1994		1995		1996		Total	
	N	%	*N*	%	*N*	%	*N*	%	*N*	%	*N*	%	*N*	%
Approach														
East	37	88.10	46	54.12	50	41.67	124	67.76	84	45.41	54	60.67	395	56.11
Mid	1	2.38	8	9.41	10	8.33	18	9.84	75	40.54	31	34.83	143	20.31
West	4	9.52	29	34.12	60	50.00	38	20.77	26	14.05	4	4.49	161	22.87
Unknown			2	2.35			3	1.64					5	0.71
Spill passage [a]														
Flood gate							30	17.05			18	21.18	48	6.82
Tainter gate	18	42.86	1	100			28	15.91			2	2.35	49	6.96
Total	*18*	*42.86*	*1*	*100*			*58*	*32.96*			*20*	*23.53*	*97*	*13.78*
Nonspill passage [b]														
Sluice	9	37.50	5	38.5 [c]	34	28.33							48	8.21
Fishpipe			14	16.47	26	21.67	56	47.46	41	23.70	35	53.85	172	29.40
Units 1–4									18	10.40	3	4.62	21	3.59
Units 9–10	13	54.17	57	67.06	57	47.50	27	22.88	42	24.28	13	20.00	209	35.73
West tube							31	26.27	68	39.31	13	20.00	112	19.15
Unknown	2	8.33	8	9.41			4	3.39	4	2.31	1	1.54	19	3.25
Fishladder			1	1.18	3	2.50							4	0.68
Total	*24*	*100*	*85*	*100*	*120*	*100*	*118*	*100*	*173*	*100*	*65*	*100*	*585*	*100*
Total detected	**42**		**85**		**120**		**183**		**185**		**89**		**704** [d]	
Mean residency (D:H:M)	0:08:09		0:21:25		0:17:37		0:08:22		0:12:00		0:06:26			

[a] Percentages are of all passages.

[b] Percentages are of all non-spill passages.

[c] Sluice closed after 13 fish had passed.

[d] Of the 704 fish detected, passage was not confirmed for all; thus, the apparent disparity in totals.

By the spring of 1991 the fishpipe was in place and operating. It was thought that with this addition, smolts that entered the forebay and traversed along the intakes toward the western high flow area would be exposed to and pass through the fishpipe.

During spring 1991, 85 (84.3% of Vernon released) emigrating radio-tagged smolts were detected at the Project. Of these, 16 had been released for a separate study at Wilder Project (121 km upstream of Vernon). The dam spilled early in the study; only one smolt passed a tainter gate during this period. The sluice was open for the first three releases ($N = 16$) and was then closed to limit passage routes to the fishpipe and turbines. This action allowed a more thorough investigation of the effectiveness of the fishpipe. Passage through the sluice in 1991 was 38.5% of all smolts detected during the period it was open. Overall, only 16.5% of all smolts monitored utilized the fishpipe; 67.1% passed through turbines.

These values were discouraging, but may have been biased by the relatively nonuniform procedures: multiple release sites, closure of the sluice gate, and so on. Mean residency time of smolts was 21 h 25 min. Upon entering the forebay, they remained there until they passed through the dam. Many smolts were detected near the trashrack for substantial time periods. Some moved back and forth along the trashracks and were detected near other intakes, but the majority of the time they were recorded near intakes 9 and 10. The substantial time periods that most smolts were detected near the intakes may suggest a surface preference of migrating salmon smolts and a general reluctance of smolts to follow flows deep (7 m to the turbine intake) in the water column, since the detection antennas at the intake areas were configured to detect tagged smolts no deeper than 2 m. Eventually, however, most did move deep and pass through turbines.

The age-2 smolts tagged and released in 1992 exhibited poor downstream movement. These were replaced with age-1 smolts from White River Hatchery, which did move downstream consistently. A total of 120 radio-tagged smolts were detected at Vernon Station in 1992 (Table 1). Half of all smolts monitored approached the dam along the western shoreline of the river. Just under 42% approached the dam from the eastern shoreline and 8.3% approached the dam from mid-river. This contrasted with both the 1990 and 1991 studies.

The reason for this disparity is not understood, although it was suspected that the approach routes contributed to the passage results. Some 47.5% of tagged smolts monitored passed Vernon Station through turbines. The fishpipe was utilized by 21.7% of the tagged smolts and the log sluice was used by 28.3%. Three smolts (2.5%) passed via the fishladder. Half the smolts entered the inner forebay from the west shoreline and were not exposed to the log sluice or the sonic deterrence system employed during this year. Once in the forebay, their only passage routes were turbines or the fishpipe. The sonic deterrence system did not exclude those smolts that approached from the eastern and mid river from the inner forebay. It was tested by periods of ensonification and no ensonification. Approximately equal numbers of tagged smolts were exposed to the on and off cycles with no detectable difference in movement and behavior between the two groups.

The results of the first three years of studies prompted NEP to investigate further measures to enhance expeditious, safe, downstream passage of anadromous fishes. After extensive flow modeling at Alden Laboratory, a louver system was designed and installed at Vernon Project during 1993. The louver was designed to intercept those fish approaching the Project from eastern and middle portions of the river and to guide them to the entrance of the fishpipe. In addition, a fishtube was installed in the westernmost portion of the inner forebay to allow passage for fish that approached down the west shoreline and for those that passed through or under the louver into the western forebay. The following three years of studies investigated the effectiveness of these two additions. The log sluice was closed for these studies to limit passage routes to only the fishpipe, fishtube, turbines, and spill gates, if high flows necessitated spillage.

1994 study

During 1994, 148 smolts were tagged and released. Of these, 139 were detected in the vicinity of the Station. An additional 44 radio-tagged smolts released for a separate study at Wilder Station were also detected at Vernon. Overall, approach to the Station was predominately from the eastern shoreline and mid-river. Mean residency times were 8 h 22 min. Of the 176 radio-tagged smolts which passed Vernon Station, 28 and 30 passed through open tainter and underwater flood gates, respectively. The fishpipe passed 47.5% of all smolts that migrated into the forebay. The western fishtube passed another 26.3% and turbines passed 22.9%. The passage route for 4 radio-tagged smolts was not determined.

Turbine passage of emigrating smolts appeared to be reduced by the presence of the louver system and west fishtube. The louver diverted up to 54.5% of the smolts that approached the dam from east- and mid-river and migrated westerly into the forebay. The remainder migrated beyond the louver into the western forebay area. Overall, 63 smolts were detected in this corner area of the forebay; 5 eventually moved to and passed via the fishpipe, 31 exited through the west fishtube, and 27 passed through turbines. This represents a 53.4% reduction in potential turbine passage due to the western fishtube. Comparison of passage data from the previous three studies indicates that turbine passage during 1994 was 50–75% less than previous years.

1995 study

The 1995 study progressed in much the same manner as the previous year, except for two modifications of the existing bypass systems. In an attempt to alter currents in front of the western fishtube to better attract smolts which tended to remain in front of the Unit 10 intake, a flow diversion panel was installed to effectively change the entrance of the passage device from approximately 7.6 m from the intake trashrack to approximately 5.5 m (Figure 3). Attraction flows were altered from primarily upstream of the entrance to downstream of it. The other addition was installation of a bar rack at the entrance to the fishpipe. This rack was installed, not as an enhancement for fish passage, but to prevent blockage of the pipe by logs and debris.

The movement and behavior of 185 radio-tagged smolts were monitored at the Station during 1995; 51 released for the Bellows Falls study were detected at Vernon after passage at that Station, 51.5 km upstream.

Approach by radio-tagged smolts to the Project was predominately from the eastern shoreline (45.4%) and mid-river (40.5%). Overall mean residency times at Vernon was 12 h.

A total of 173 radio-tagged smolts passed Vernon Station in 1995. Of these, 41 (23.7%) passed through the fishpipe, 68 (39.3%) passed via the west fishtube, and 60 (34.7%) passed through turbines. Passage routes for 4 (2.3%) radio-tagged smolts were not determined. Of all turbine-passed smolts, 18 passed through Units 1–4, and 42 passed via Units 9 and 10.

The louver system excluded from the western forebay up to 42.1% of those smolts which approached the dam from east- and mid-river. This apparent decrease of approximately 12% from the 1994 study may have been due to unusually low river flows in 1995; frequently during the study, the powerhouse was at 50% or less capacity. This low flow severely altered the hydraulics of the louver; it was modeled to work most efficiently at maximum generation. Additionally, fishpipe passage for those smolts apparently diverted by the louver was not complete or expeditious. For all smolts which never entered the west forebay, 16.4% were not detected passing the Station during the monitoring period, 6.0% passed by unknown routes, 25.4% passed through Units 1–4, and 52.2% passed via the fishpipe. Those that passed through the fishpipe resided in the eastern forebay area for a mean time of 12 h before passing. The increased mean residency time compared to the 1994 study, and passage through Units 1 through 4, which never occurred in previous year's studies, was most likely due to the fishpipe entrance trashrack. This rack was quite frequently clogged with debris, limiting access. A total of 117 radio-tagged smolts (excluding western-approaching smolts) migrated beyond the louver system into the western forebay area. Of these, 6 migrated back and exited via the fishpipe, 1 migrated back and passed through one of Units 1–4, 1 passed via Unit 9, 41 exited through Unit 10, and 68 passed through the west fishtube. Fishtube passage represented 61.8% of all western forebay passage. In 1994, without the flow diverter, 53.4% of all smolts that passed from the west forebay did so through the fishtube. This suggests that the installation of the fishtube flow diverter board may have increased fishtube passage.

1996 study

For the 1996 study, the plan was identical to the previous year's study except for two changes. The trash bar rack in front of the fishpipe entrance was removed and a turbine operation cycling schedule was established. Turbines 10, 9, 7, 6, and 1–4 were to be put on line in that order, and taken off line in reverse order. This schedule was intended to maximize the flow field along the louver for better efficiency.

Commencement of the 1996 study was in jeopardy, however, due to unusually high spring rainfall and elevated river flows that necessitated high spillage rates. Since a high proportion of emigrating smolts pass through open spill gates, and the focus of the investigation was the effectiveness of the louver system, release of smolts had to be postponed until spill decreased significantly. This, coupled with a flood gate failure (gate could not be closed), prompted discussions between NEP and the CRASC on the viability of proceeding with the study; upon mutual consent, the study proceeded. The tag and release of salmon was delayed appreciably. The first group of fish was not released until 30 May (relatively late in the season) at a water temperature of 14.5°C. The remaining five groups were released between 1 and 8 June. Of the 147 tagged fish released, 89 (60.5%) were detected at Vernon Station. The delayed release of salmon and their questionable condition (many salmon exhibited parr marks) probably contributed to the relatively small number of fish detected at the Station.

Approach to the Project of the 89 radio-tagged fish was predominately from the eastern shoreline (60.7%) and mid-river (34.8%). Overall, mean residency time at the Station was 6 h 26 min. Of the 85 salmon which were detected passing the Station, 20 (23.5% of all passage detections) passed through spill gates, 35 (41.2%) passed the fishpipe, 13 (15.3%) passed via the west fishtube and 16 (18.8%) passed through turbines. The passage route for 1 (1.2%) radio-tagged fish was not determined and 4 salmon detected did not pass the Station. Of 16 turbine passed salmon, 3 passed by Units 1–4, and 13 passed via Unit 10.

The louver system diverted up to 62.9% of the smolts that approached the dam from the east and mid-river. This is an increase from the 42.1% noted for 1995 and 54.5% in 1994. Fishpipe utilization for all nonspill passages also increased from 23.7% in 1995 to 53.9% in 1996. Passage time was substantially less and passage through Units 1–4 decreased from 10.4% in 1995 to 4.6% in 1996. These data are indicative that removal of the fishpipe trashrack and initiation of a turbine dispatch protocol may have facilitated passage through the fishpipe.

Twenty-three salmon migrated beyond the louver system into the western forebay area. Three more approached down the west shoreline and directly entered the west forebay. Of these 26 fish, 13 passed through Unit 10 and 13 passed via the fishtube.

Bellows Falls project

Between 3 and 16 May 1995, 152 radio-tagged smolts were released upstream of the Station in six groups (Table 2). River water temperature during the study period ranged from 10.5 to 13.5°C. Station discharge ranged from 34 to 298 m^3/s.

Overall, 144 smolts (94.7%) were detected at the Station. Times for smolts to reach the Station after release varied from 4 min to 23 h 14 min and averaged (mean) 59 min. The Station was generating at full capacity upon arrival for all smolts in groups 1, 2, and 6; generation varied from 1 unit to full capacity during arrival for smolts in groups 3, 4, and 5.

The fish diversion boom excluded up to 121 (84.0%) tagged smolts from the area of the Station intakes (inner forebay). Residency times in areas upstream of the boom averaged 11 min; times ranged from less than 1 min to 1 h 47 min. All but one passed the Station via the sluice; one smolt was not detected passing. In total, 83 tagged smolts excluded by the boom passed the Station in less than 5 min; 34 more passed within 1 h and 4 passed between 1 and 2 h

Twenty-three (16.0%) smolts passed under the boom into the inner forebay. Residency times in the inner forebay ranged from 1 min to 1 d 9 h 18 min; the mean was 4 h 58 min. Eight of these fish passed within 10 min; 4 by sluice and 4 by turbine 3. Three additional smolts passed within 1 h; 2 by turbine and 1 by sluice. Eleven of the remaining 12 smolts passed the

Table 2. Data summary for all radio-tagged Atlantic salmon smolts released upstream of Bellows Falls Station, Spring 1995. (E = Excluded from forebay; P = Passed into forebay)

	Release group 1		2		3		4		5		6		Total	
Date	3 May		5 May		10 May		12 May		14 May		16 May			
Time	11:44		10:35		11:30		11:50		14:50		10:10			
H_2O temp (°C)	10.5		11.0		12.0		12.5		13.5		13.0			
Size (mm FL)														
Low	165		164		172		173		167		178		164	
Mean	186.0		182.8		185.7		186.7		189.0		190.0		186.8	
High	218		210		210		206		210		205		218	
Released	27		25		25		25		25		25		152	
Detected	25		24		25		23		23		24		144	
Passed	**25**		**24**		**24**		**23**		**23**		**24**		**143**	
Generation[a] (cfs)														
Low	9,900		10,200		6,700		1,200		6,700		9,800		1,200	
Mean	10,056		10,200		9,324		8,883		7,126		9,896		9,271	
High	10,500		10,200		9,800		10,200		10,200		9,900		10,500	
Generation[b] (cfs)														
Low	9,800		9,600		3,100		1,200		3,300		9,800		1,200	
Mean	10,108		10,167		8,800		8,961		7,178		9,896		9,207	
High	10,500		10,400		9,900		10,200		10,200		9,900		10,500	
Diversion boom	E	P	E	P	E	P	E	P	E	P	E	P	E	P
Number	13	12	22	2	22	3	22	1	19	4	23	1	121	23
Residency time (H:M)														
Low	0:01	0:01	0:00	2:19	0:00	8:20	0:00		0:00	4:31	0:00		0:00	0:01
Mean	0:08	1:43	0:31	2:34	0:04	13:11	0:04	0:03	0:15	12:11	0:03	0:07	0:11	4:58
Median	0:02	0:11	0:32	2:34	0:02	12:06	0:01	0:03	0:05	5:27	0:01	0:07	0:02	2:19
High	0:55	8:39	1:47	2:48	0:50	19:06	0:42		1:13	1:09:18	0:25		1:47	1:09:18
Passage route														
Sluice	13	6	22	1	21	2	22	1	19	4	23	1	120	15
Turbines 1-2		0		0		0		0		0		0	0	0
Turbine 3		6		1		1		0		0		0	0	8

[a] At time of arrival.
[b] At time of passage.

Station within 24 h after arrival and 1 smolt resided in the inner forebay for greater than 24 h before passing. In total, 15 smolts entered the secondary access gate and passed via the sluice; the remaining 8 passed through turbine 3.

Overall, 135 of the tagged smolts (94.4% of those passing) were detected and passed Bellows Falls Station by the sluice. Eight (5.6%) passed via turbine 3. One smolt was detected at the Station, but was not detected passing.

Smolts may have approached the Station in groups. Times to the Station (time from release to first detection) for many individuals within the same release groups were identical. Once at the Station behavior of smolts differed; some held upstream of the diversion boom longer than others and some went under the diversion boom into the inner forebay. The reason for this is not clear but may be due to individual behavior characteristics or due to the rapid and diverging currents in the vicinity of the boom. Surface water is diverted along the face of the boom to the sluice, while deeper water, at a much greater volume, is pulled under the boom to the generators. The vertical point, or depth, along the boom at which water is diverted under is unknown and probably variable along the length of the structure.

Movement of the 38 smolts that took greater time to pass was generally either a circling pattern around the upstream area of the boom and a holding pattern in the vicinity of the sluice entrance, before passing. Some smolts were monitored in the corner area just upstream of the sluice entrance for up to 25 min before they passed. Others were detected on all diversion boom antennas multiple times indicating both clockwise and counter-clockwise circular movements immediately upstream of the boom. One smolt spent 50 min upstream of the diversion boom before vanishing. This smolt was not detected on any monitoring station after this period and presumably moved back upstream.

Relatively few (1–4) smolts from each release group, except group 1, entered the inner forebay. The disproportionate number of fish (12 of 25 monitored) from group 1 which entered the inner forebay could not be explained by Station generation, time of day, and water temperature at times of arrival. These variables, for group 1, were very similar, and in some cases identical, to those for other groups. The 23 smolts that entered the inner forebay under the boom exhibited varied movement patterns before and after entering. Some immediately went under the boom while others moved along the upstream side of the boom, both with and against the current, for periods of up to 1 h before they passed under into the inner forebay. Either end of the diversion boom served as the access points where smolts entered the inner forebay. No smolts were detected moving under the boom in the middle sections, though some could have entered these sections rapidly and not been detected on the boom antennas. Some smolts were detected on the boom main antenna and then at the intake, indicating immediate movement under the boom. Those smolts that entered at the western portion usually had moved extensively along the boom for some length of time; those that moved under at the eastern portion typically were monitored holding in that area (near the sluice gate).

Smolts that passed under the boom into the inner forebay spent from 1 min to more than 24 h before passing. Movement of these fish was varied. Some immediately passed the Station, either by turbine or sluice, while others held position near Unit 3 intake before passing. Still others exhibited extensive movement about the inner forebay for extended periods before passing. These smolts were detected multiple times on the intake antennas, and periodically at the diversion boom (backside of antennas). The relative strength of signals received at the boom in conjunction with near immediate detection at the intake indicates that these fish were confined within the inner forebay, downstream of the boom. No smolts, once in the inner forebay, passed to the upstream side of the boom. All smolts, regardless of initial detection near the intakes (a few were initially detected near both Units 1 and 2), eventually were drawn to the corner area of Unit 3 intake and the secondary access gate to the sluice; surface flow appeared to be predominately toward this corner, even during full generation. All smolts that entered the inner forebay either passed the Station via the secondary access gate to the sluice ($N = 15$) or Unit 3 ($N = 8$).

The effect of flow, or Station generation, on the efficiency of the diversion boom to guide emigrating smolts to the sluice is not readily discernible due to insufficient data at the lower range of intake flow. All groups of tagged smolts, except group 5, were released during full generation (272–289 m^3/s) at the Station; group 5 was released with two Units generating (198 m^3/s). Although a wide range of flows was not tested, full generation can arguably be considered the worst case condition. The probability of fish being drawn under the diversion boom should increase with increasing volume of water flowing under the boom to the turbines. All but four smolts which entered the inner forebay did so at times of full generation; the four entered during two units (2 and 3) operating. Excluding group-5 fish, eight smolts approached the Station at less than full generation. Flows varied from 34 to 190 m^3/s. All eight smolts were diverted by the boom and passed through the sluice; all but two passed in less than 5 min, one passed in 11 min and the other in 42 min.

A total of 51 smolts that passed Bellows Falls Station was detected and monitored at Vernon Station. Of these, 48 and 3 passed Bellows Falls via sluice and turbine, respectively.

Conclusions

The Vernon Station bypass of emigrating salmon smolts generally improved with the addition of surface water passage systems and subsequent fine-tuning of the systems. The log sluice, by itself, was found to be insufficient in passing an adequate proportion of smolts by the Station. Adding the fishpipe and its operation in conjunction with the log sluice apparently improved bypass effectiveness some 12%; when operated by itself, however, the fishpipe performed relatively poorly. The attraction flow of the fishpipe apparently was not sufficient to compete with the strong currents flowing primarily toward Units 9 and 10 in the western forebay area. Apparently this flow tended to draw and concentrate high numbers of smolts. Installation of the louver system seemed to facilitate interception and guidance of emigrating salmon to the fishpipe bypass. Passage through the pipe increased twofold from the previous study year. Addition of the alternate bypass (western fishtube) further increased bypass efficiency. In the final year of study, after turbine cycling protocol was initiated, bypass efficiency was 73.9%, up from 37.5% in 1990. Passage survival studies conducted on Units 10 and 4 in the spring of 1996 resulted in survival rates of 94.9% for Unit 10 and 81.0% for Unit 4 (NAI 1996). Atlantic salmon smolt survival studies were also conducted on the west fishtube during 1995 and resulted in 93.3% passage survival (NAI 1995). These values in combination with bypass study results suggest that 95.5% of emigrating salmon were afforded safe passage by Vernon Station. The CRASC accepted these results and requested no further studies or improvements be conducted.

The fish diversion boom at Bellows Falls Station was successful in facilitating the expeditious passage of tagged salmon smolts. In total, 84% of all detected smolts that approached the Station did not enter the inner forebay and were diverted to the bypass sluice entrance. Of these, 69% passed in 5 min or less; the remainder passed in less than 2 h. The secondary access gate to the sluice, within the inner forebay, offered a viable alternate route for those smolts that were not excluded. Of these, 65% entered the access entrance and passed the Station via the sluice. The remainder passed through Unit 3. Overall, 94.4% of all detected Station passages were via the sluice.

These results were obtained under an assumed worst-case condition with respect to prevailing flows. Five of the six release groups were released during periods of full generation so that most smolts, upon arrival at the Station, were exposed to full generation. Only a small proportion (16%), however, utilized turbines for passage. It is likely that guidance efficiency may be higher at lower flows.

These results, in conjunction with a sluice passage survival study conducted in 1991 (RMC 1991) showing a 96.0% survival rate, met the criteria described in the Memorandum of Agreement. The CRASC accepted these results, and even though at least three years of study were initially required, did not request further study or improvements.

References

RMC (RMC Environmental Services, Inc.). 1991. Survival of Atlantic salmon smolts passing the log-ice sluice at Bellows Falls Hydroelectric Station, Vermont. Report of RMC prepared for New England Power Company, Westborough, Massachusetts.

NAI (Normandeau Associates, Inc.). 1995. The Vernon bypass fishtube: evaluation of survival and injuries of Atlantic salmon smolts. Report of NAI prepared for New England Power Company, Westborough, Massachusetts.

NAI (Normandeau Associates, Inc.). 1996. Estimation of survival and injuries of Atlantic salmon smolts in passage through two Francis turbines at the Vernon Hydroelectric Station, Connecticut River, Vermont. Report of NAI prepared for New England Power Company, Westborough, Massachusetts.

Stephen V. Amaral, Edward P. Taft, Frederick C. Winchell, Anthony Plizga, Edward Paolini, and Charles W. Sullivan

Fish Diversion Effectiveness of a Modular Inclined Screen System

4

A new type of fish diversion screen, known as the modular inclined screen (MIS), has been designed to provide effective fish protection at any type of water intake. Because the screen operates at water velocities of up to about 3 m/s in the approach channel, the MIS is more compact and cost-effective than existing low-velocity screens. The biological effectiveness of the MIS was evaluated in laboratory tests conducted with eleven fish species and in a field evaluation conducted with six fish species. During the laboratory tests, MIS passage survival exceeded 99% at velocities up to 1.8 m/s for most species, although lower survival was observed for the relatively fragile blueback herring at high operating velocities. The field evaluation demonstrated similar trends in passage survival related to species and velocity. Blueback herring passage success decreased considerably at velocities greater than 1.2 m/s, whereas passage of several riverine species typically exceeded 90% at velocities up to 1.8 m/s. The laboratory and field evaluations have demonstrated that the MIS has potential to successfully divert a wide range of species at water intakes.

Introduction

Water withdrawals for electric generating facilities and irrigation systems can negatively impact resident, anadromous, and catadromous fish populations. In an attempt to mitigate these adverse impacts, safe diversion and bypass of fish is often a regulatory requirement at water intakes throughout the United States. Resource agencies and private industry have expended considerable effort and resources to develop biologically effective technologies (EPRI 1986, 1994a; INEL 1997). The application of many technologies is limited because they are designed for use at a specific site or type of intake, or with certain species and size classes. Additionally, many fish protection systems that are considered effective or are accepted for use by resource and regulatory agencies often cannot be economically applied.

The Electric Power Research Institute (EPRI) developed a high-velocity modular inclined screen (referred to as the MIS; U.S. Patent No. 5385428) to address the need for a fish protection system that has the potential for application at a variety of water intakes and with a wide range of species and size classes. The MIS also was designed to be more cost-effective than most low-velocity fish screen systems currently in use (EPRI 1994b). The modular design of the MIS allows for one or more units to be installed at most types of water intakes (e.g., hydroelectric, cooling water, and irrigation diversion) over a wide range of intake flow volumes.

A laboratory testing program was developed to evaluate the hydraulic performance (ARL 1993; Cook et al. 1993) and biological effectiveness of the MIS. In laboratory studies, fish diversion effectiveness was examined at five module intake water velocities using 11 fish species (both riverine and anadromous fishes). Based on the effective and safe passage of fish observed during the laboratory tests, a field evaluation of a prototype MIS was conducted with several riverine species and the anadromous blueback herring *Alosa aestivalis* at the Green Island Hydroelectric Project on the Hudson River in 1995 and 1996.

Modular Inclined Screen Design

The MIS consists of an entrance with trashracks, dewatering stop log slots, a wedge-wire screen set at an angle to the flow between 10 and 20°, and a bypass for guiding diverted fish to a transport pipe or channel (Figure 1). The screen is composed of 50% porosity bar with 1.9 mm spacing and is mounted on a pivot shaft so that it can be cleaned by backflushing. The screen in a full-scale module would be 9.1 m in length by 3.0 m in width, with an effective area of about 23.2 m^2 and the capacity to screen up to 2.8 m^3/s at 3.0 m/s. The module is completely enclosed and is designed to operate at water velocities ranging from about 0.6–3.0 m/s, depending on species and life stages to be protected and project intake flows.

Before the biological evaluation, the original MIS design was refined during hydraulic studies conducted with a 1:6.6 scale model (Cook et al. 1993). Results of the hydraulic model tests demonstrated that the MIS entrance design created an acceptable velocity distribution, even when approach flows were skewed as much as 45°. The modular design features were effective in developing uniform velocities over the screen surface without any high velocity zones.

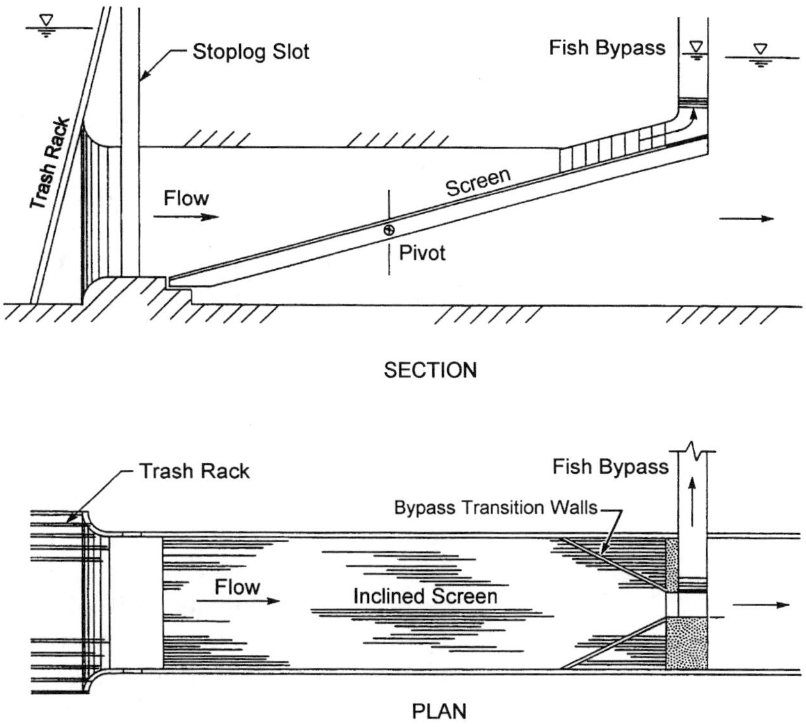

Figure 1. Schematic of modular inclined screen (MIS).

Several bypass geometries were evaluated during the hydraulic studies. A configuration that appeared to offer the greatest potential for effective and safe fish passage was selected for the biological test model. Additionally, bypass flows were regulated such that the water velocity at the bypass entrance was equal to the module water velocity. Using the refined design developed during the hydraulic model studies, a 1:3.3 scale model was constructed for the laboratory biological evaluation.

Methods

Laboratory evaluation

The laboratory evaluation of MIS diversion effectiveness was conducted in 1992 and 1993. Fish species evaluated in laboratory tests were selected to represent the types of fish that are of greatest concern at water intakes in the United States. The species that were evaluated included bluegill *Lepomis macrochirus*, walleye *Stitzostedion vitreum*, channel catfish *Ictalurus punctatus*, blueback herring, American shad *Alosa sapidissima*, rainbow trout *Oncorhynchus mykiss*, coho salmon *O. kisutch*, chinook salmon *O. tshawytscha*, golden shiners *Notemigonus crysoleucas*, brown trout *Salmo trutta*, and Atlantic salmon *Salmo salar* smolts. Blueback herring and American shad were collected during the fall out-migration from the Connecticut River and were tested as one group (hereafter referred to as clupeids). All other species were obtained from a private distributor of hatchery-reared fish. The mean fork length of fish that were evaluated was less than 100 mm for each species, except Atlantic salmon, which averaged about 170 mm in length (Table 1). Two size classes of rainbow trout were evaluated; one group was classified as fry (51 mm mean fork length; FL) and the other was classified as juveniles (66 mm mean FL).

Diversion effectiveness was evaluated for each species at module intake water velocities of 0.6, 1.2, 1.8, 2.4, and 3.0 m/s. Between one and eight trials were conducted at each module velocity (Table 2). With the exception of clupeids, each species was evaluated at all five test velocities (Table 2). Target sample sizes ranged from 25 to 100 fish for treatment and control groups, depending on the number of fish that were available (the total number of treatment and control fish released for each species tested is presented in Table 2). The actual number of fish released often was less than the target sample size because fish that had illegible marks, visible injuries, or were exhibiting abnormal swimming behavior were removed from treatment and control groups before release. Control groups were used to assess injury and latent mortality associated with testing procedures. In 1992, paired releases of treatment and control fish were performed for most tests. Controls were not used during bluegill and walleye tests that were conducted with reduced handling techniques (reduced handling tests were performed with these species after high and variable latent mortality was observed during initial tests). In 1993, one control group usually was released for tests performed at each velocity (i.e., one control group per multiple treatment groups per module velocity).

Colored ink marks or fin clips were used to distinguish between treatment and control groups and fish released at different module velocities. A pressurized air injection system was used to release treatment fish into the test flume approximately 0.3 m upstream of the center of the MIS entrance. Less than 2,100 kg/m^2 was required for displacing the water and fish from the injection pipe into the flume. After passing through the MIS model, treatment fish exited the bypass into a fish collection area that had a net pen (1.5 m wide by 1.8 m long by 1.2 m deep) placed below the bypass outfall weir. Control groups were released into the net pen immediately after treatment fish were released. The module screen was examined to determine the number and location of fish that were impinged after each test was completed.

Table 1. Mean fork lengths (standard deviations are in parentheses) for sub-samples of fish evaluated during the MIS laboratory tests. Treatment and control fish always were collected from the same lots of fish held in holding pools.

Species	Treatment fish		Control fish	
	N	Mean FL (mm)	*N*	Mean FL (mm)
Bluegill	104	47 (6)	140	47 (6)
Walleye	63	85 (4)	72	87 (6)
Channel catfish	139	88 (12)	146	89 (10)
Rainbow trout fry	149	47 (5)	175	49 (5)
Rainbow trout juvenile	90	52 (5)	102	51 (5)
Clupeids (1992)	54	76 (5)	72	74 (5)
Clupeids (1993)	936	67 (9)	280	67 (9)
Golden shiner	75	77 (7)	75	76 (7)
Brown trout	50	59 (6)	50	60 (5)
Coho salmon	75	49 (5)	75	50 (4)
Chinook salmon	75	54 (4)	75	53 (3)
Atlantic salmon	48	170 (15)	25	169 (14)

Table 2. Number of tests conducted per velocity and number of fish released and recovered for each species evaluated during MIS laboratory tests.

Species	Module velocities tested (m/s)	Tests per species	Treatment releases: Total released	Treatment releases: Total recovered	Control releases: Total released	Control releases: Total recovered
		1992 tests				
Bluegill	0.6, 1.2, 1.8, 2.4	3	556	481	570	565
Bluegill	0.6, 1.2, 1.8	2	300	288		
(reduced handling)	2.4, 3.0	1	104	102		
Walleye	0.6, 1.2, 1.8, 2.4	3	285	247	281	281
Walleye	0.6, 1.2, 1.8	2	300	279		
(reduced handling)	2.4, 3.0	1	165	165		
Channel catfish	0.6, 1.2, 1.8, 2.4, 3.0	3	761	743	736	732
Rainbow trout fry	0.6, 1.2, 1.8, 2.4, 3.0	3	644	522	616	612
Rainbow trout juvenile	0.6, 1.2, 1.8, 2.4, 3.0	3	566	554	486	486
Clupeids	0.6	8	200	183	200	199
	1.2, 1.8, 2.4	4	300	295	300	297
		1993 tests				
Clupeids	0.6, 1.2, 1.8, 2.4	6	1,048	962	313	280
Golden shiner	0.6, 1.2, 1.8, 2.4, 3.0	3	1,046	946	355	355
Brown trout	0.6, 1.2, 1.8, 2.4, 3.0	3	750	632	243	242
Coho salmon	0.6, 1.2, 1.8, 3.0	3	588	475	194	194
	2.4	6	297	290	49	49
Chinook salmon	0.6, 1.2, 1.8	3	457	433	154	154
	2.4, 3.0	6	753	751	309	463
Atlantic salmon	0.6, 1.2, 1.8, 2.4, 3.0	3	450	291	400	381

Treatment and control fish were dip-netted from the bypass collection pen and were examined for scale loss and visible injuries and measured for fork length. Scale loss was recorded in increments of 10% for the side with the most loss. For most tests, fish collected live were held for a 72 h to assess latent mortality (48-h holding periods were used for clupeid tests conducted at module velocities of 0.6 and 2.4 m/s in 1993). Reduced handling tests with bluegill and walleye were conducted in the same manner as standard tests, but treatment fish were not marked and were examined for injury and measured for length after the latent mortality holding period ended. Similarly, to reduce stress from handling, clupeids and Atlantic salmon smolts were not evaluated for injury or measured for length until the end of the latent mortality period.

Fish diversion data were used to calculate three biological effectiveness parameters for each combination of species, module water velocity, and test condition (i.e., clean screen and debris accumulation) that were evaluated. These measures included:

- the percent of fish diverted live;
- MIS passage survival (based on survival from time of release to the end of the latent mortality holding period); and
- diversion-related injury.

Data from replicate tests were combined to produce pooled estimates of each parameter for each species and test condition that was evaluated.

The percent of fish diverted live (which is equivalent to a measure of immediate screen passage survival) and MIS passage survival were calculated using the relative recovery rate method for paired release–recapture experiments described in Burnham et al. (1987). This method has been used in similar studies that have estimated survival relative to treatment effects of a passage or bypass route (i.e., turbine or spillway passage) based on the release and recovery of marked treatment and control groups (Mathur et al. 1994, 1996). The relative recovery rate method calculates passage survival by dividing the survival rate of treatment fish by the survival rate of control fish. MIS passage survival was estimated using immediate and latent survival rates. When more than one trial was conducted for a set of test conditions (i.e., velocity and species), the data were combined to generate pooled estimates of survival (Burnham et al. 1987). Confidence intervals were calculated for each estimate of percent diverted live and MIS passage survival using maximum-likelihood methods described by Burnham et al. (1987).

Injury rates attributed to MIS passage were estimated by dividing the rate of uninjured treatment fish by the rate of uninjured control fish and subtracting the result from one. Fish were considered injured if there were any visible external injuries or if there was scale loss greater than three percent.

An assessment of the effects of organic debris accumulation on screen hydraulics and fish diversion also was performed. Debris tests were conducted with rainbow trout, golden shiner, and coho, chinook, and Atlantic salmon. Diversion of rainbow trout during debris tests was assessed using pine needles (ponderosa and Jeffrey pines, 100–200 mm in length), deciduous leaves (white birch, black oak, and maple species), and aquatic vegetation (*Vallisneria* species); each of these debris types were evaluated separately (Table 3). Tests with golden shiner and the three species of salmon all were conducted with deciduous leaves.

Debris tests were conducted at module velocities up to 2.4 m/s (Table 3) and with up to seven levels of debris-related head loss per velocity. The total number of treatment fish released per species and debris type tested ranged from 190 (rainbow trout) to 917 (golden shiner; Table 3). Control groups were not released during debris tests due to fish availability constraints. Control data from tests with a clean screen were used to estimate MIS passage survival and injury rates for each species tested with debris.

Prototype Field Evaluation

A field evaluation of a prototype MIS was conducted at the Green Island Hydroelectric Project in 1995 and 1996. The Green Island Project is owned and operated by Niagara Mohawk Power Corporation and is on the Hudson River in Troy, New York. An MIS test facility (Figure 2) was installed upstream of an existing ice sluice gate that is between the project's auxiliary dam and forebay skimmer wall.

Table 3. Number of treatment fish released and recovered for each species evaluated with debris accumulation on the MIS during laboratory tests. Control groups were not released during debris tests (control data from clean screen tests with the same species and module velocities were used to calculate MIS passage survival for debris tests).

Species	Debris type	Module velocities tested (m/s)	Total fish released	Total fish recovered
Rainbow trout fry	pine needles	0.4, 1.2, 1.8, 2.4	819	683
	deciduous leaves	1.8	127	110
	aquatic vegetation	1.8	190	165
Coho salmon	deciduous leaves	0.6, 1.2, 1.8, 2.4	402	376
Chinook salmon	deciduous leaves	0.6, 1.2, 1.8, 2.4	901	880
Golden shiner	deciduous leaves	0.6, 1.2, 1.8, 2.4	917	588
Atlantic salmon	deciduous leaves	0.6, 1.2, 1.8, 2.4	225	181

The fish passage area of the screen facility had a total length of approximately 9.1 m, a width of 1.5 m and a height of 1.2 m. The entrance to the MIS had a trashrack with 20-cm spacings between bars. The screen was approximately 1.5 m wide by 4.9 m long. Transition walls adjacent to the downstream portion of the screen guided fish into the bypass entrance, which was 0.3 × 0.3 m in cross-section. Bypass flows were discharged into a fish collection area and were regulated by adjusting the height of a bottom-drop gate at the terminus of the bypass. A collection net, which sampled the entire flow passing through the MIS, was positioned downstream of the screen and was used to recover debris and impinged fish after the screen was backflushed.

Modular inclined screen tests were conducted between late September and early November in 1995 and 1996. Species that were evaluated in 1995 included blueback herring (juvenile outmigrants), rainbow trout, and golden shiner. Largemouth bass *Micropterus salmoides*, smallmouth bass *M. dolomieui*, bluegill, and yellow perch *Perca flavescens* were evaluated in 1996. Largemouth and smallmouth bass were tested as one group. Blueback herring passage was evaluated with fish that entered the MIS as they encountered the project during their out-migration (i.e., naturally entrained), and with fish that were collected from the project forebay using a lift net. The other species all were obtained from hatchery sources. The mean fork length of each species evaluated during field tests was less than 100 mm (Table 4).

Tests with each species were conducted at several screen approach velocities between 0.6 m/s and 2.4 m/s. Between one and six tests were conducted at each module velocity (Table 5). Groups of about 50–100 test fish were marked with a fin-clip and injected into the MIS just upstream of the screen using a pressurized release system similar to the one used during the laboratory evaluation. Control fish were released into the bypass flume immediately after the introduction of treatment fish. The treatment and control fish were recovered together in a hopper at the outfall to the bypass flume. Collected fish were transferred to holding tanks where they were observed for 48 h to determine latent mortality. At the end of

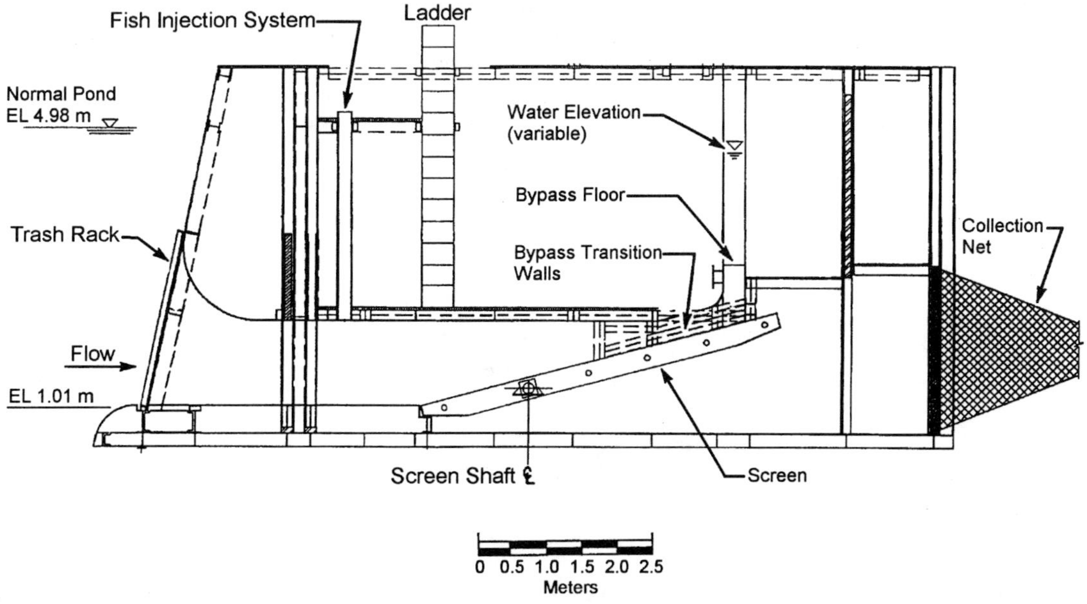

Figure 2. Green Island modular inclined screen (MIS) test facility.

Table 4. Length data for species evaluated during the MIS field tests.

Species	*N*	Mean FL (mm)	Range (mm)
Blueback herring (naturally entrained)	620	55.9	42–93
Blueback herring (injected)	1,661	65.5	41–94
Rainbow trout	836	94.7	54–122
Golden shiner	1,109	70.5	43–95
Largemouth/ smallmouth bass	1,620	69.4	41–120
Bluegill	1,677	70.6	34–130
Yellow perch	1,253	60.0	44–130

each test, the screen was backwashed and the downstream collection net was checked for impinged fish. The same biological effectiveness parameters that were estimated for laboratory tests also were used to assess the performance of the prototype MIS tested at Green Island.

Results

Laboratory evaluation

A total of 20,739 fish was tested during the biological evaluation of the MIS in 1992 and 1993; 16,314 during clean screen tests and 4,425 during debris tests. Of 9,045 treatment fish that were recovered during clean screen tests, 97.3% were collected live, 0.3% were collected dead, and 2.4% were classified as impingements (i.e., any fish that was impinged on the screen

Table 5. Number of tests conducted per velocity and number of fish released and recovered for each species evaluated during the MIS Green Island field evaluation. The number of blueback herring released during injection tests is estimated; exact counts were not conducted to minimize stress from handling. Consequently, the number of fish recovered during these tests may be more than the number released.

			Treatment releases		Control releases	
Species	Module velocities tested (m/s)	Tests per species	Total released	Total recovered	Total released	Total recovered
		1995 tests				
Blueback herring	0.6	1		101		
(naturally entrained)	1.2	2		519		
Blueback herring	0.6	1	100	105		
(injected)	1.1	3	180	126	100	77
	1.2, 1.8	6	>1,050	1,193	350	325
Rainbow trout	1.2, 1.8, 2.4	3	1,350	1,179	450	44
Golden shiner	0.6, 1.2, 1.8, 2.3	3	1,200	1,068	600	588
		1996 tests				
Largemouth and smallmouth bass	0.6, 1.2, 1.8, 2.3	3	900	796	899	835
Bluegill	0.6, 1.2, 1.8, 2.3	3	900	857	900	862
Yellow perch	0.6, 1.2, 1.8, 2.3	3	776	698	774	686

at the end of a test). Of the 6,022 control fish that were recovered, 99.8% were collected live and 0.2% were collected dead. Of 2,967 test fish that were recovered during debris tests, 93.9% were collected live, none were collected dead, and 6.1% were impinged.

A significant modification was made to the screen in 1993 when a 1.3-cm wide strip of duct tape was applied to the screen edges along the transition walls leading to the bypass. Before the modification, approximately 82.5% of all recorded impingements occurred along the transition wall screen edges. No impingements were observed on these screen edges after the screen was modified. All data for species evaluated in 1993 are for tests that were performed after the screen was modified. This feature has been incorporated into the basic MIS design by physically blocking the underside of the wedge-wire screen along a narrow strip extending 1.3 cm out from the transition walls.

With the exception of clupeids, the percent of fish diverted live exceeded 95% at all module velocities for each species evaluated in 1992, and was greater than 96% at all velocities for species tested in 1993 (Table 6).

For most species, MIS passage survival was slightly lower than percent diverted live due to latent mortality (Table 7). MIS net passage survival typically exceeded 92% at module velocities of up to and including 2.44 m/s for species tested in 1992 and exceeded 93% at all five velocities for species tested in 1993 (Table 7). Atlantic salmon smolts had a passage survival rate of 100% at all five test velocities.

Treatment fish injury rates were low and often comparable to the injury rates of control fish, with the exception of clupeids (Table 8). Adjusted injury rates for the majority of tests conducted in 1992 and 1993 were 0% (Table 8). The majority of injuries sustained by treatment and

Table 6. Percent of treatment fish diverted live (i.e., immediate survival) and ±95% CI for each species evaluated during laboratory tests. Confidence intervals could not be calculated for reduced-handling tests with bluegill and walleye because control groups were not used.

	Module intake velocity (m/s)				
Species	0.6	1.2	1.8	2.4	3.0
Bluegill	97.6 ± 15.4	99.2 ± 3.2	100.0 ± 0.0	97.8 ± 5.2	not tested
Bluegill (reduced handling)	100.0	99.0	99.0	98.0	96.2
Walleye	100.0 ± 0.0	100.0 ± 0.0	98.5 ± 6.8	91.7 ± 16.9	not tested
Walleye (reduced handling)	100.0	100.0	99.0	95.2	100.0
Channel catfish	99.2 ± 2.8	100.0 ± 0.0	100.0 ± 0.0	99.4 ± 2.6	98.6±5.7
Rainbow trout fry	98.5 ± 5.8	100.0 ± 0.0	100.0 ± 0.0	95.2 ± 10.5	95.6±9.3
Rainbow trout juvenile	100.0 ± 0.0	100.0 ± 0.0	100.0 ± 0.0	99.1 ± 4.5	96.3±8.5
Clupeids (1992)	89.1 ± 4.5	97.0 ± 3.3	81.6 ± 7.7	63.5 ± 9.6	not tested
Clupeids (1993)	99.8 ± 0.5	95.8 ± 3.1	94.2 ± 3.5	94.7 ± 2.9	not tested
Golden shiner	99.1 ± 1.7	100.0 ± 0.0	99.1 ± 1.2	98.6 ± 1.6	98.4±1.8
Brown trout	97.9 ± 2.8	100.0 ± 0.0	100.0 ± 0.0	98.8 ± 1.7	98.6±2.0
Coho salmon	100.0 ± 0.0	100.0 ± 0.0	100.0 ± 0.0	99.0 ± 1.2	100.0±0.0
Chinook salmon	100.0 ± 0.0	100.0 ± 0.0	100.0 ± 0.0	99.7 ± 0.7	96.3±1.7
Atlantic salmon	100.0 ± 0.0	100.0 ± 0.0	100.0 ± 0.0	100.0 ± 0.0	100.0±0.0

control fish were scale loss greater than 3%. The rates of treatment fish with scale loss greater than 3% were minimal (less than 5%) for most species. Additionally, after factoring in control fish scale loss, the estimated scale loss rate due screen passage was even lower.

Live diversion and MIS passage survival of juvenile clupeids was lower than observed for the other species tested. The percent of clupeids diverted live ranged from a high of 97.0% at a module velocity of 1.2 m/s to a low of 63.5% at 2.4 m/s (Table 6). Clupeids were not tested at 3.0 m/s due to high impingement rates at the lower velocities. Clupeids sustained higher rates of latent mortality than the other species that were tested, and, subsequently, net passage survival also was lower (Table 7). Latent mortality, however, was considerable for both treatment and control fish during clupeid tests (greater than 30% at all velocities).

The percent of clupeids diverted live improved in 1993 after the screen transition wall edges were blocked (Table 6). The screen modification considerably reduced the number of fish that were impinged at all test velocities. Despite the increased live diversion rate, MIS passage survival of clupeids did not demonstrate a corresponding increase. Clupeid passage survival was similar between the two test years (Table 7).

Injury rates of clupeid treatment and control groups were excessive, exceeding 80% for all four velocities that were evaluated with clupeids. Injury rates attributed to MIS passage, which include scale loss estimates, were 0% at 1.2, 1.8, and 2.4 m/s and only 2.3% at 0.6 m/s (Table 8). This demonstrates that nearly all observed injuries were not due to MIS passage, but rather handling associated with collection and testing.

Table 7. MIS passage survival and ±95% CI for each species and module velocity evaluated during laboratory tests. Confidence intervals could not be calculated for reduced-handling tests with bluegill and walleye because control groups were not used. Values greater than 100 percent indicate control fish mortality was greater then treatment fish mortality.

Species	Module intake velocity (m/s)				
	0.6	1.2	1.8	2.4	3.0
Bluegill	85.9 ± 51.2	102.2 ± 29.7	90.8 ± 35.9	78.0 ± 18.6	not tested
Bluegill (reduced handling)	100.0	99.0	97.0	98.0	94.3
Walleye	95.8 ± 9.7	89.4 ± 54.2	86.2 ± 35.0	90.8 ± 43.7	not tested
Walleye (reduced handling)	98.7	100.0	99.0	95.2	100.0
Channel catfish	99.2 ± 2.8	100.0 ± 0.0	100.0 ± 0.0	99.4 ± 2.6	98.6±5.7
Rainbow trout fry	92.6 ± 18.1	100.0 ± 0.0	103.5 ± 8.6	95.2 ± 24.2	91.2±17.8
Rainbow trout juveniles	100.7 ± 10.2	99.2 ± 3.0	101.0 ± 3.1	99.4 ± 2.4	89.9±11.9
Clupeids (1992)	77.9 ± 20.8	91.9 ± 20.4	74.2 ± 17.9	74.3 ± 64.4	not tested
Clupeids (1993)	76.1 ± 7.2	24.0 ± 17.9	74.8 ± 87.2	69.1 ± 12.3	not tested
Golden shiner	99.1 ± 1.7	98.7 ± 3.4	95.1 ± 4.0	91.9 ± 3.7	94.7±3.2
Brown trout	93.6 ± 6.7	100.0 ± 0.0	99.8 ± 4.9	98.8 ± 1.7	104.0±7.6
Coho salmon	100.0 ± 0.0	100.0 ± 0.0	100.0 ± 0.0	101.0 ± 4.2	99.3±1.3
Chinook salmon	100.0 ± 0.0	102.0 ± 4.0	99.3 ± 1.3	98.0 ± 1.6	93.8±2.7
Atlantic salmon	100.0 ± 0.0	100.0 ± 0.0	100.0 ± 0.0	101.4 ± 2.8	100.0±0.0

Table 8. Injury rates (%) attributed to MIS passage for each species and module velocity evaluated during laboratory tests. The majority of injuries observed were scale loss greater than three percent. Passage injury rates for bluegill and walleye tests with reduced handling techniques were not calculated because control groups were not released during these tests.

	Module intake velocity (m/s)				
Species	0.6	1.2	1.8	2.4	3.0
Bluegill	1.6	0.8	4.5	2.2	not tested
Walleye	0.0	0.0	1.5	0.0	not tested
Channel catfish	0.8	0.0	0.0	0.0	0.7
Rainbow trout fry	1.4	0.0	0.9	0.8	2.0
Rainbow trout juveniles	1.7	1.6	0.0	0.0	1.9
Clupeids (1992)	2.3	0.0	0.0	0.0	not tested
Clupeids (1993)	0.0	0.0	0.0	0.0	not tested
Golden shiner	2.7	0.1	4.5	21.5	37.4
Brown trout	3.1	0.0	0.0	0.6	0.7
Coho salmon	0.0	0.0	0.7	0.0	0.0
Chinook salmon	0.0	0.0	0.0	0.0	2.0
Atlantic salmon	0.0	18.9	0.0	0.0	0.0

Debris tests demonstrated a clear relationship between the weight of debris on the screen and head loss across the screen (Figure 3). Incremental increases in head loss associated with debris weight were greatest for deciduous leaves and lowest for pine needles. The relationship between debris-related head loss and passage survival varied with species and module velocity, but the general trend was a reduction in passage survival with increases in head loss. Observed reductions in passage survival were the direct result of increases in impingement rates.

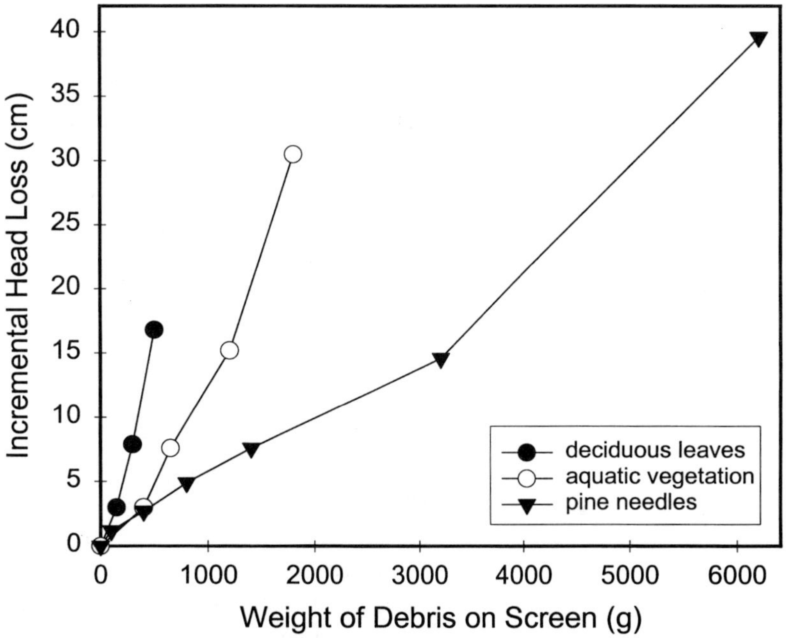

Figure 3. Incremental head loss across the modular inclined screen (MIS) due to accumulation of three types of debris.

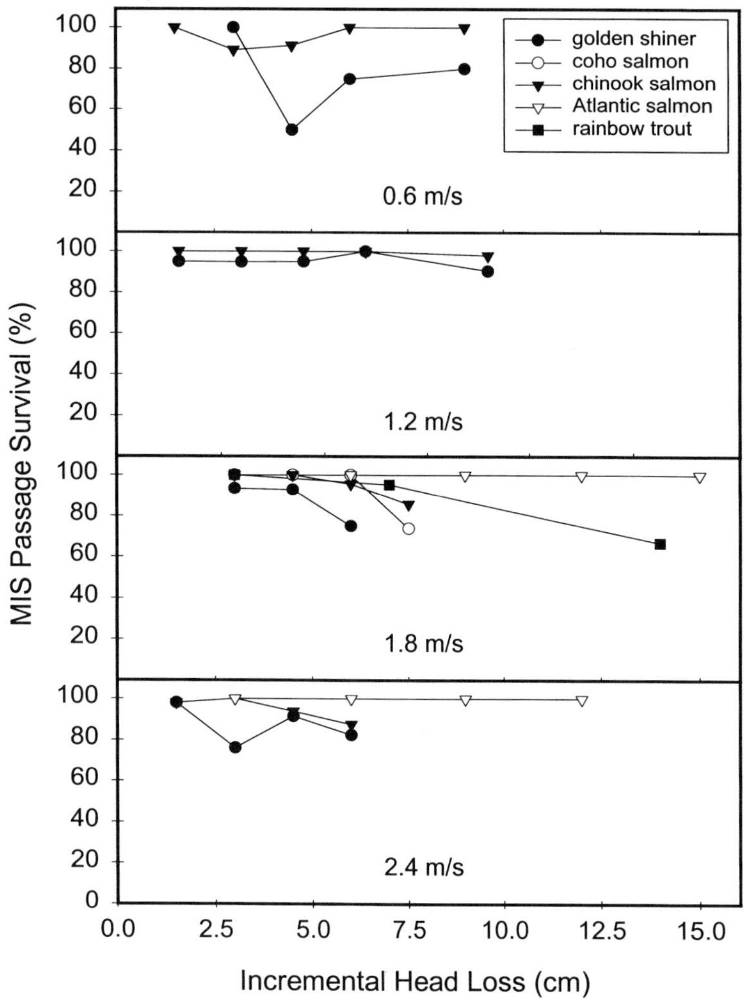

Figure 4. Modular inclined screen (MIS) passage survival of five species evaluated during tests with deciduous leaf debris on the screen.

Of the five species evaluated with deciduous leaf debris, golden shiner appeared the most susceptible to decreases in passage survival, with reduced survival as head loss increased at three of the four velocities that were evaluated (Figure 4). Chinook salmon passage survival was not affected by increases in head loss at the two lowest velocities (0.6 and 1.2 m/s), but decreased at head losses greater than 5 cm at a velocity of 1.8 m/s and at head losses greater than 2.5 cm at 2.4 m/s (Figure 4). Coho salmon and rainbow trout, evaluated only at a velocity 1.8 with deciduous leaf debris, demonstrated decreases in passage survival as head losses exceeded 5 cm (Figure 4). Atlantic salmon passage survival was unaffected by the presence of debris on the screen at the two velocities tested with this species (Figure 4).

Rainbow trout exhibited decreased passage survival with increases in head loss at all four of the velocities that were evaluated with pine needle debris (Figure 5). Decreases in MIS passage survival were more pronounced as velocity increased, with the lowest survival rates occurring at an approach velocity of 2.4 m/s. Similar to tests with deciduous leaves, passage survival began to decline at each velocity when screen

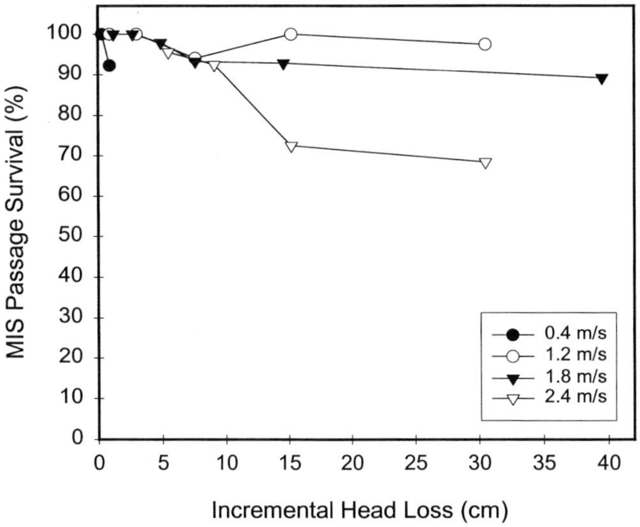

Figure 5. Modular inclined screen (MIS) passage survival of rainbow trout during tests with pine needle debris on the screen.

head loss exceeded 5 cm (Figure 5). Passage survival of rainbow trout also appeared to be dependent on debris type. During tests with each debris type at a velocity 1.8 m/s, MIS passage survival of rainbow trout declined more rapidly as head loss increased when deciduous leaves were on the screen (Figure 6). Rainbow trout survival appeared least affected by the accumulation of pine needles.

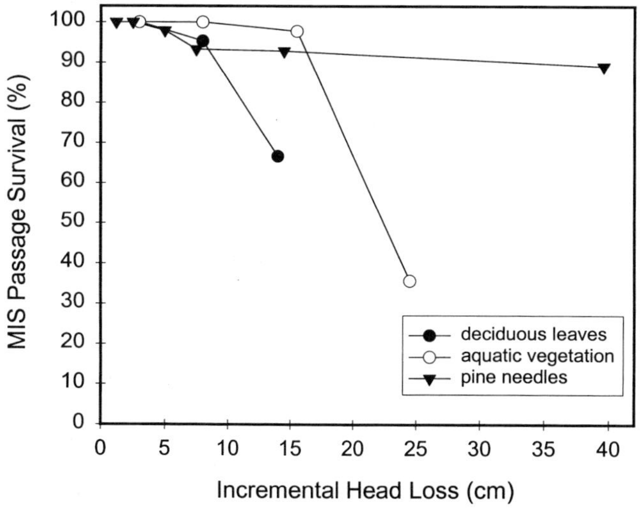

Figure 6. Modular inclined screen (MIS) passage survival of rainbow trout evaluated with three debris types at a module velocity of 1.8 m/s.

Table 9. Percent diverted live (i.e., immediate survival) and MIS passage survival (with ± 95% CI's) for species evaluated during the MIS field tests. Values greater than 100 percent indicate that the mortality rate for control fish was greater than it was for treatment fish.

Species/year of test	MIS diversion parameter	Module intake velocity (m/s)			
		0.6	1.2	1.8	2.4
Rainbow trout (1995)	diverted live	not tested	100.0 ± 0.0	100.0 ± 0.0	99.3 ± 2.9
	passage survival	not tested	100.0 ± 0.0	100.0 ± 0.0	99.3 ± 2.9
Golden shiner (1995)	diverted live	99.2 ± 3.8	100.0 ± 0.0	99.3 ± 3.2	93.5 ± 25.3
	passage survival	97.2 ± 12.2	101.6 ± 11.0	99.2 ± 9.0	90.9 ± 22.9
Largemouth/ smallmouth bass (1996)	diverted live	100.0 ± 0.0	98.5 ± 3.7	96.1 ± 7.0	99.1 ± 1.8
	passage survival	100.0 ± 3.6	98.5 ± 3.7	96.1 ± 7.0	97.7 ± 4.7
Bluegill (1996)	diverted live	100.0 ± 0.0	99.0 ± 2.0	91.6 ± 9.3	99.4 ± 2.8
	passage survival	99.4 ± 2.9	100.0 ± 3.6	88.5 ± 4.4	89.5 ± 17.1
Yellow perch (1996)	diverted live	97.9 ± 0.7	98.8 ± 3.0	76.0 ± 35.1	54.1 ± 51.7
	passage survival	98.5 ± 2.5	95.2 ± 12.3	74.5 ± 30.6	54.8 ± 50.7

Prototype field evaluation

Biological test results at Green Island in 1995 and 1996 were similar to those obtained during the laboratory evaluation. Live diversion rates and MIS passage survival often exceeded 95% for tests with riverine fishes (Table 9). The percent diverted live and passage survival for blueback herring (naturally-entrained and injected; Tables 10 and 11) were lower than observed for the other species tested. However, with the exception of one test at 1.2 m/s, the live diversion rates of injected blueback herring were greater than 9%. Based on the laboratory and field tests with injected fish, if control groups could have been used during natural entrainment tests to account for handling injury and mortality at the time of collection, the percent diverted live and passage survival most likely would have been higher.

Bluegill, golden shiner, and rainbow trout were diverted live at rates that were comparable to those estimated for each of these species during laboratory tests. MIS passage survival of golden shiner at the Green Island facility was similar to the laboratory results up to a velocity of 1.8 m/s. Passage

Table 10. Diversion parameter estimates for MIS field tests with naturally-entrained blueback herring. MIS passage survival is derived from immediate and latent mortality rates. Immediate mortality included fish that were recovered dead from the bypass and fish that were classified as impingements. The estimates of each of these parameters are considered conservative because they are not adjusted for mortality and injury associated with handling (i.e., control groups were not used for these tests), which can be excessive for juvenile clupeids.

Module intake velocity (m/s)	Test date	Percent diverted live	MIS passage survival (%)	Percent with scale loss >3%
0.6	9 Oct	98.1	88.3	40.4
1.2	3 Oct	89.7	58.7	87.1
1.2	19 Oct	57.7	19.7	87.0

Table 11. Diversion parameter estimates (with ±95% CI's) for MIS field tests with blueback herring that were injected into the screen entrance. Control data from the 15 October test at 1.1 m/s was used to calculate the diversion parameters for the 17 October test at 0.6 m/s. Injury rates reflect the incidence of scale loss greater than 3%.

Module intake velocity (m/s)	Test date	Percent diverted live	MIS passage survival (%)	MIS passage injury rate (%)
0.6	17 Oct	91.7 ± 6.2	85.9 ± 13.5	0.0
1.1	15 Oct	96.5 ± 4.5	78.3 ± 12.7	67.4
1.2	17 Oct	95.4 ± 2.4	74.6 ± 14.5	37.8
1.2	19 Oct	94.0 ± 3.8	54.6 ± 15.6	75.7
1.8	16 Oct	61.4 ± 4.5	33.3 ± 5.8	82.4
1.8	18 Oct	93.9 ± 3.5	3.9 ± 2.5	98.2

survival of bluegill was approximately 10% lower during field tests than it was in laboratory tests at velocities of 1.8 and 2.4 m/s. Diversion and survival rates for rainbow trout during the field and laboratory evaluations almost were identical, exceeding 99% at most velocities.

The rate of fish with scale loss associated with MIS passage generally increased with velocity and was high (greater than 10%) for golden shiner, largemouth and smallmouth bass, and bluegill at a velocity of 2.4 m/s (Table 12). Passage-related scale loss also was excessive for bluegill at a velocity of 2.8 m/s.

The evaluation of naturally-entrained blueback herring was limited due to low numbers of fish entering the MIS facility. However, with the use of the fish injection system, substantial data on the effectiveness of the MIS was gathered with blueback herring captured from the forebay. Despite undesirable fish handling that was required with this system, the diversion efficiencies and MIS passage survival rates were similar to laboratory results.

Diversion and passage survival rates for blueback herring that were naturally entrained into the MIS facility were highest at the lowest of the two velocities tested (0.6 and 1.2 m/s; Table 10). Additionally, the rate of scale loss was considerably lower at 0.6 m/s (Table 10). Relative differences in survival and injury rates between tests at the two velocities are indicative of the effects of a higher approach velocity on passage success

Table 12. Rate of fish with scale loss (greater than three percent on one side of fish) attributed to MIS passage during the field tests in 1995. Negative values indicate that the rate of scale loss for control fish was greater than it was for treatment fish.

Species, year tested	Module intake velocity (m/s)			
	0.6	1.2	1.8	2.4
Rainbow trout (1995)		5.7	−0.9	3.6
Golden shiner (1995)	0.0	4.0	4.6	44.4
Largemouth/ smallmouth bass (1996)	0.0	1.1	0.5	11.4
Bluegill (1996)	0.1	3.5	48.8	69.1
Yellow perch (1996)	−0.6	0.6	3.6	4.0

of blueback herring. However, the estimates of passage survival and scale loss for naturally-entrained blueback herring are not truly representative of passage success because control data were not available to account for mortality and injury associated with collection handling and latent mortality holding.

Live diversion rates for injected blueback herring were adjusted with control data and exceeded 90% for all tests, except for one conducted at 1.8 m/s (Table 11). MIS passage survival of injected blueback herring declined with increases in velocity (Table 11). The rate of scale loss greater than 3% also was related to velocity, with the lowest rate of scale loss occurring at 0.6 m/s and the highest rate at the 1.8 m/s.

Discussion

The results of the MIS biological evaluation (laboratory and field tests) demonstrated that the MIS has the potential to effectively and safely divert a wide range of fish species. Live diversion rates and MIS passage survival of riverine fishes and anadromous salmonids typically were between 95 and 100%, even at screen approach velocities as high as 3 m/s. Additionally, injury attributed to MIS passage was low, rarely exceeding 5%. During field tests, notable exceptions to high passage survival rates for riverine species were bluegill and yellow perch at module velocities of 1.8 and 2.4 m/s. Additionally, bluegill suffered extensive scale loss rates at these velocities. Blueback herring diversion and survival rates were lower than observed for the other species tested, and injury rates were higher. However, high mortality and injury rates experienced by both treatment and control fish make the results from blueback herring tests less reliable and more difficult to interpret. Specifically, control fish mortality greater than 10% can compromise the accuracy of passage survival rates (Ruggles 1992).

The results of the Green Island field evaluation also provided valuable information on design, fabrication, installation, and operational considerations (EPRI 1996; ARL 1996) that indicate the concept has good potential for application at other hydroelectric projects and water intakes. Mechanically, the MIS performed well; screen seals were adequate for preventing fish escapement downstream while allowing free movement of the screen during backwashing. Backwashing was performed on numerous occasions with complete removal of debris. After each backflushing, the screen reseated properly when raised back to the normal operating position. With its submerged inlet, the MIS design largely eliminates concerns for entrainment of larger floating debris.

Although the MIS has not been used in a full-scale application to date, the development and application of the Eicher screen has demonstrated the efficacy of using inclined screens with high approach velocities to protect fish at water intakes (EPRI 1992, 1994b; Smith 1993, 1997; Cramer 1997). The similar design principle, the common developmental approach, and the biological effectiveness of the Eicher screen indicate that the MIS has potential to be equally effective. Similar to the MIS, the Eicher screen is a passive pressure screen that operates at high velocities. An extensive evaluation of an Eicher screen was conducted with juvenile salmonids at the Elwha Hydroelectric Project in Washington. All species evaluated were diverted efficiently and safely by the Eicher screen to a bypass pipe over the 1.2–2.4 m/s penstock velocity range; fish were successfully diverted at rates exceeding 98% for most size classes that were evaluated (EPRI 1992). The first full-scale Eicher screen installation (two screens in two, 3-m diameter penstocks; total flow of 28 m^3/s) at B. C. Hydro's Puntledge Project produced similar results, with passage survival rates of chinook and coho salmon smolts exceeding 99% at penstock velocities up to 1.8 m/s (Smith 1993, 1997).

Advantages of the MIS over conventional low-velocity screen systems include lower installation costs and potential reductions in passage delays. The size of screening facilities is highly dependent on approach velocity criteria (Pearce and Lee 1991). This means costs will be higher for facilities with lower velocity criteria. The ability of fish to locate bypasses in a timely manner also is an important issue for diversion systems (Rainey 1985). Once fish have entered an MIS facility, they are generally swept quickly over the screen and into a bypass (actual transport times will be dependent on fish species and size and approach velocity). Major concerns associated with the use of high-velocity screens include descaling from screen contact and impingement. The evaluations of both the MIS and the Eicher screen have demonstrated that, for most species tested, rates of descaling and impingement are negligible (less than 5%).

The laboratory and field evaluations have demonstrated that the MIS has considerable potential to safely divert a wide range of species at water intakes. The next step in the development of the MIS as a viable fish protection alternative will be the installation and evaluation of a full-scale facility. As is true with the application of any fish protection technology, careful consideration of engineering parameters (project design and hydraulic conditions) and biological criteria (species and size) will be necessary to select a site where the MIS will have a high probability of success.

References

ARL (Alden Research Laboratory, Inc.). 1993. Hydraulics of a new modular fish diversion screen. Prepared for Stone & Webster Engineering Corporation, Boston.

ARL (Alden Research Laboratory, Inc.). 1996. 1996 hydraulic testing of the modular inclined screen at the Green Island test facility. Prepared for Stone & Webster Engineering Corporation, Niagara Mohawk Power Corporation, and New York State Energy Research and Development Authority, Boston.

Burnham, K. P., D. R. Anderson, G. C. White, C. Brownie, and K. H. Pollock. 1987. Design and analysis methods for fish survival experiments based on release-recapture. American Fisheries Society, Monograph 5, Bethesda, Maryland.

Cook, T. C., E. P. Taft, G. E. Hecker, and C. W. Sullivan. 1993. Hydraulics of a new modular fish diversion screen. Pages 318–326 *in* W. D. Hall, editor. Waterpower '93: proceedings of the international conference on hydropower. American Society of Civil Engineers, New York.

Cramer, D. P. 1997. Evaluation of a louver guidance system and Eicher screen for fish protection at the T. W. Sullian Plant in Oregon. Proceedings of the fish passage workshop. Section 2. Alden Research Laboratory, Inc., Conte Anadromous Fish Research Center, Electric Power Research Institute, and Wisconsin Electric Power Company. Alden Research Laboratory, Holden, Massachusetts.

EPRI (Electric Power Research Institute). 1986. Assessment of downstream migrant fish protection technologies for hydroelectric application. EPRI Report 2694-1, Palo Alto, California.

EPRI (Electric Power Research Institute). 1992. Evaluation of the Eicher screen at Elwha Dam: 1990 and 1991 test results. EPRI Report TR-101704, Palo Alto, California.

EPRI (Electric Power Research Institute). 1994a. Research update on fish protection technologies for water intakes. EPRI Report TR-104122, Palo Alto, California.

EPRI (Electric Power Research Institute). 1994b. Fish protection/passage technologies evaluated by EPRI and guidelines for their application. EPRI Report TR-104120, Palo Alto, California.

EPRI (Electric Power Research Institute). 1996. Evaluation of the modular inclined screen (MIS) at the Green Island hydroelectric project: 1995 test results. EPRI, Palo Alto, California.

INEL (Idaho National Engineering Laboratory). 1997. Hydropower research and development. Prepared for the U.S. Department of Energy, DOE/ID-10575, Idaho Falls, Idaho.

Mathur, D., P. G. Heisey, and D. A. Robinson. 1994. Turbine-passage mortality of juvenile American shad at a low-head hydroelectric dam. Transactions of the American Fisheries Society 123:108–111.

Mathur, D., P. G. Heisey, E. T. Euston, J. Skalski, and S. Hays. Turbine passage survival estimation for chinook salmon smolts (*Oncorhynchus tshawytscha*) at large dam on the Columbia River. Canadian Journal of Fisheries and Aquatic Sciences 53:542–549.

Pearce, R. O., and R. T. Lee. 1991. Some design considerations for approach velocities at juvenile salmonid screening facilities. Pages 237–248 *in* J. Colt and R. J. White, editors. Fisheries bioengineering symposium. American Fisheries Society, Symposium 10, Bethesda, Maryland.

Rainey, W. S. 1985. Consideration in the design of juvenile bypass systems. Pages 261–268 *in* F. W. Olson, R. G. White, and R. H. Hamre, editors. Symposium on small hydropower and fisheries. American Fisheries Society, Western Division and Bioengineering Section, Bethesda, Maryland.

Ruggles, C. P. 1992. Effect of stress on turbine fish passage mortality estimates. Canadian Technical Report of Fisheries and Aquatic Sciences 1905:39–57.

Smith, H. A. 1993. Development of a fish passage solution at the Puntledge hydro intake facility. Pages 197-204 *in* K. Bates, compiler. Fish passage policy and technology. American Fisheries Society, Bioengineering Section, Bethesda, Maryland.

Smith, H. A. 1997. Operating history of the Puntledge River Eicher screen facility. Proceedings of the fish passage workshop. Section 3. Alden Research Laboratory, Inc., Conte Anadromous Fish Research Center, Electric Power Research Institute, and Wisconsin Electric Power Company. Alden Research Laboratory, Inc., Holden, Massachusetts.

Charles M. Peven and Thaddeus R. Mosey

Development of Surface Bypass and Collection at Rocky Reach Dam, Columbia River

5

Between 1985 and 1994, Chelan County Public Utility District (Chelan) used rotating and passive diversion screens in the turbine intakes of Rocky Reach Dam to guide fish away from the turbines and around the dam. Guidance efficiency of the diversion screen system was insufficient, rarely exceeding 25% of the total fish entering the turbine intakes. As a result, Chelan pursued the concept of surface collection based primarily on the success of spillway baffles at Wells Dam. Chelan tested a prototype surface collector in 1995 to see if the concept was feasible at Rocky Reach. Approximately 900,000 juvenile fish passed through the prototype. Chelan concluded the concept of surface collection at Rocky Reach was feasible. In 1996, we extended the floor of the surface collector upstream and added a sloping wall. Hydraulic and mathematical modeling showed that the new sloping wall and floor extension negatively affected the direction of water flow into the entrance of the collector. We removed the extended floor and sloping wall before the 1997 field season. In 1996, we observed approximately 30% of the radio-tagged steelhead that entered the entrance of the collector proceeded all the way through. In 1997, the percentage increased to 73%. The passive integrated transponder tag study in 1996 showed that approximately 25% of the juvenile chinook and steelhead were guided by the fish bypass system. In 1997, the percentage improved to 47%. The hybrid fish passage system (surface collector and diversion screens) appears to be the preferred method for juvenile salmonid bypass at Rocky Reach Dam in the future.

Introduction

The Rocky Reach Hydro Project is in mid-Washington State on the Columbia River, about 11 km upstream from the city of Wenatchee at river kilometer (RKM) 762.7 (Figure 1). The powerhouse runs parallel to the river flow, while the spillway runs perpendicular (Figure 2). Rocky Reach Dam is owned and operated by Chelan County Public Utility District (Chelan).

In the early 1980s, Chelan began investigations into the development of a bypass system for juvenile salmonids *Oncorhynchus* spp., which would divert downstream migrating smolts away from the turbine intakes and safely pass them around the dam to the tailrace. Rotating and passive diversion screens were tested in the intakes of the dam from 1985 to 1994 and found to be ineffective in guiding fish (Peven and Keesee 1992; Peven and Abbott 1994). Due to the low guidance efficiencies (less than 25%), none of the configurations tested were developed into a final bypass system. Although tests of the diversion screens in Turbine Unit 1 were deemed unsuccessful, Chelan left these screens in Unit 1 to provide some means of interim protection to fish that use the system as the final bypass system is developed. Evaluation of the screens after the surface collector was installed suggests that the fish guidance efficiency of the screens increased.

As a result of the poor prototype screen performance, Chelan decided to investigate the concept of surface bypass and collection, based on the success of spillway baffles at Wells Dam (RKM 830.2; Kudera and Sullivan 1993). The feasibility of this bypass concept was further supported by the efficiency of fish passage through trash sluiceways at various dams on the Snake and Columbia rivers (Sverdrup Corporation, unpublished data, 1993).

Hydroacoustic studies at Rocky Reach Dam have shown that most fish approaching the powerhouse travel in the upper 15.2 m of the water column, pass by most of the upstream turbine units, and enter the most downstream units (Raemhild et al. 1984; Steig and Sullivan 1991). Over 70% of juvenile

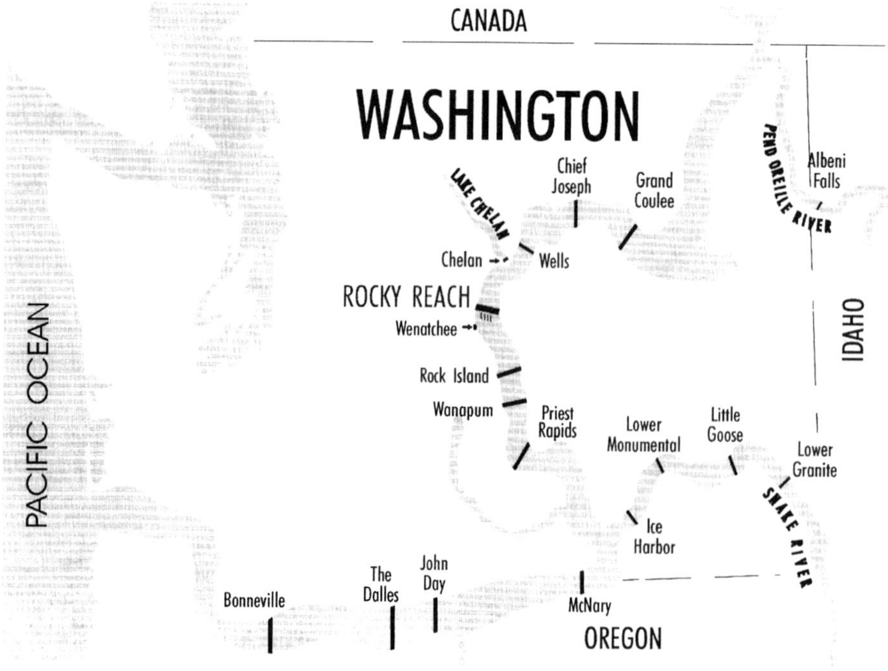

Figure 1. Location of Rocky Reach Dam on the Columbia River.

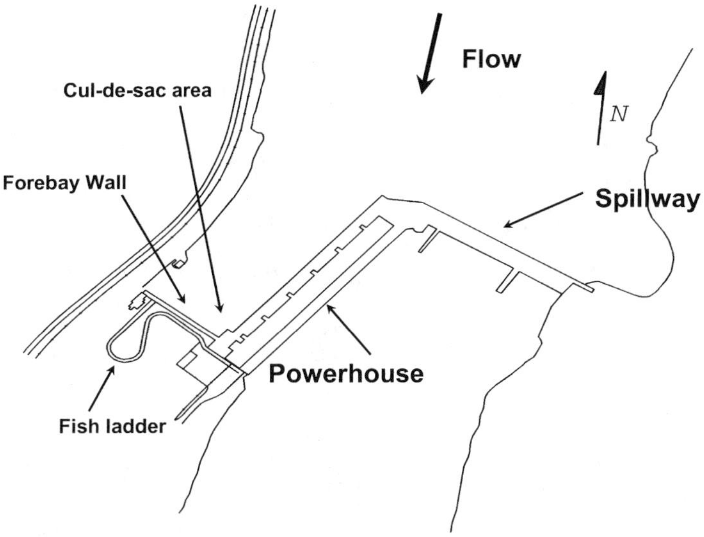

Figure 2. Diagram of Rocky Reach Dam showing major components of the project.

salmonid migrants pass through the turbines at the downstream end of the powerhouse (Units 1–3; Figure 3). Hydroacoustic surveys have also shown that many fish appear to circulate within the forebay area near the powerhouse in front of Units 1 and 2, and the service bay without immediately sounding to enter a turbine intake (Steig et al. 1995).

Chelan's approach to the surface collection concept is based on the forebay hydraulic characteristics and the concentration of fish in the forebay near Units 1–3. Chelan, with the help of a hydraulic model of the forebay (see below), positioned the opening of the prototype within forebay surface currents (upper 18.3 m) believed to contain substantial numbers of fish known to concentrate within the vicinity of Units 1–3.

In 1993, Chelan began constructing a 1:30 scale forebay model at ENSR Consulting and Engineering in Redmond, Washington to obtain information about flow characteristics in the forebay of Rocky Reach. By spring 1994, a surface collection prototype structure was developed. Construction began in fall 1994, and the prototype was first tested in 1995.

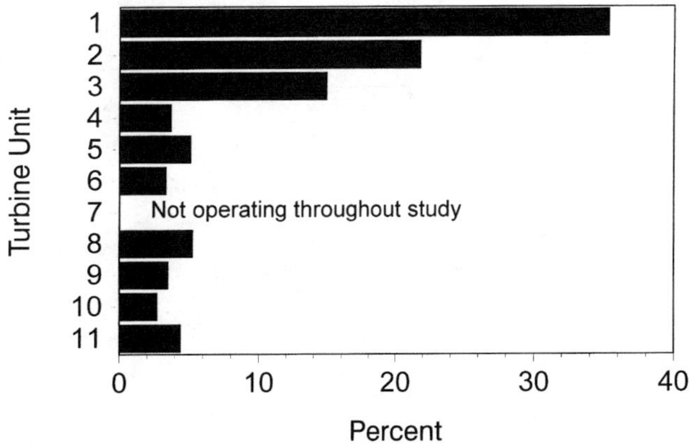

Figure 3. The horizontal distribution of fish passing Rocky Reach Dam in 1983 as estimated from hydroacoustics (Raemhild et al. 1984).

Figure 4. The 1995 surface collector configuration at Rocky Reach Dam showing the major components of the system. (Graphics courtesy of CH2M HILL.)

We evaluated the concept of whether surface collection was feasible at Rocky Reach with the first configuration of the structure in 1995 (Figure 4). We enumerated juvenile salmonids as they passed through the system, subsampled the bypassed population for species composition and fish condition (descale/injury), and made observations of fish behavior and movement with hydroacoustic monitoring (conducted by Hydroacoustic Technology, Inc., Seattle, Washington). By recording fish

Figure 5. The 1996 surface collector configuration at Rocky Reach Dam showing the extended floor and sloping wall. (Graphics courtesy of CH2M HILL.)

Figure 6. The 1997 surface collector configuration showing the major changes; primary screen module, secondary screen expansion, and new pump station. (Graphics courtesy of CH2M HILL.)

passage through the system with a video camera, we counted nearly 900,000 fish using the system (Peven et al. 1995). Although this information did not give Chelan a clear indication of fish passage efficiency, we concluded the concept was worth further development.

In an effort to improve the hydraulics of the forebay currents leading into the surface collector opening (and subsequent fish passage efficiency), we made three structural changes between 1995 and 1996 (Figure 5):

- the floor of the surface collector was extended from the opening upstream to the powerhouse wall at the point separating Units 2 and 3,
- a sloping wall was added, and
- an additional venturi was added, increasing total flow from 50.4 m^3/s to 65.8 m^3/s.

(Venturi openings are used to draw the flow through the primary screen area. The water discharges directly into Unit 1.)

The floor was added to create a sharp flow separation near the bottom of the surface collector entrance. It was postulated that fish would detect this sharp flow separation, avoid the quickly descending flow into the unit and stay in the surface currents where their chance of guidance was higher. The sloping wall was added because it appeared (in the hydraulic model) to increase the direction of flow towards the surface collector entrance. The flow volume was not increased.

Testing in 1996 suggested that the modifications that were made were deleterious to fish passage, and consequently, we made additional changes to the system for testing in 1997:

- the extended floor and sloping wall were removed;
- a pump station was added in the secondary dewatering area to pass more flow;
- the geometry of the downstream end of the collector entrance (near

the primary screens) was modified from an almost 90° (1.2 m wide) turn to a sweeping bend (1.8 m wide; Figure 6).

These changes were made to improve hydraulic conditions both within (accelerating flow instead of decelerating flow) and immediately upstream (direction of flow) of the collector's entrance and to reduce the physical constriction of the downstream end of the entrance.

In 1997, we also installed a weir and transport system attached to the headgate slots (gatewell collection system); the gatewell collection system is where fish that were deflected by the diversion screens are collected. The weir and transport system reduced the amount of handling and potential stress to fish that were transported to the tailrace.

In 1996 and 1997, in addition to the aforementioned components of our monitoring program, we added the capability to measure fish passage efficiency of the fish bypass system by installing four 0.61 m passive integrated transponder (PIT) tag detectors in the fish transport pipe. We also inserted radio telemetry tags in juvenile chinook *O. tshawytscha* and steelhead *O. mykiss* to monitor their passage and behavior patterns as they approached the project. For the purpose of this paper, we will concentrate on the results of the PIT tag releases and the radio telemetry observations.

Methods

Guidance equipment

Surface collector

The surface collector prototype system at Rocky Reach consists of two vertical guide walls that attach to two 1.2 m vertical columns and run between the forebay wall and the north end of generating Unit 1. A floor connects the two guide walls, making it a closed system. The vertical columns also form the 4.6-m-wide by 17.4-m-high entrance to the system. The velocity of water entering the surface collector at the upstream edge of the opening ranged from 0.67 m/s to 0.94 m/s. In 1995 and 1996, the maximum flow entering the surface collector entrance was approximately 66 m^3/s. Venturi devices, located in the floor of the system immediately upstream of Unit 1, draw a maximum of about 54 m^3/s through the primary dewatering system (primary screens) into Unit 1. The primary screens extend from the vertical columns to the forebay wall. The primary dewatering area floor is solid and slopes upward. The flow that passes the primary dewatering area (2.8 m^3/s in 1995 and 1996; increased to 14 m^3/s in 1997) enters the secondary dewatering system and passes fish in approximately 0.6 m^3/s into the bypass transport piping system. The transport system passes through an inverted siphon through the forebay wall near the fish ladder exit. The pipe then follows the fishway wall until it reaches a switch that shunts the flow of the pipe either to the juvenile facility (where fish are sampled) or continues on through the rest of the system. The pipe continues down the inside of the south fishway wall, and turns north, following the tailrace wall until it reaches the Unit 3 discharge, where it extends over the water, discharging the fish and water into the tailrace downstream of the turbine boil. The outfall piping system was designed to flow half full with discharge velocities of less than 9 m/s. The outfall pipe ranges from 1.22 to 4.88 m above the tailrace level, depending on the tailrace elevation.

In 1997, a steel structure was placed at the upstream edge of the entrance of the surface collector. This structure reduced the entrance opening width from 4.5 to 1.8 m; the opening remained 17.4 m deep. The purpose of installing this structure was to test whether a higher velocity opening would be more attractive to fish approaching the surface collector entrance. The structure increased the velocity from approximately 0.94 m/s to about 1.65 m/s, although the total flow volume entering the system dropped from about 68 m^3/s to 52 m^3/s. We anticipated that fish might be more prone to leave the collector after entering it (with the slot narrowing device in place) because of the flow separations that were created downstream of the opening. Considering this, we decided that the appropriate metric for measuring the effects of the narrower entrance was whether radio-tagged fish were detected within the entrance, regardless of their final destination.

Gatewell collector, Unit 1

Before the 1997 season, weirs were installed in the headgate slots of Unit 1. One 0.51 m weir was installed in the south slot (0.56 m^3/s max.), and two 0.30 m weirs in the center and north slots (0.28 m^3/s max.). The differences in the number and size of weirs enabled us to test differences in flow rates and patterns to see which weir configuration was most efficient. The slots are connected to a collection channel that passes through a dewatering device. The channel then passes through a gallery under the forebay deck, along the forebay wall to the right bank fish ladder, then parallel to the surface collector bypass pipe. The channel passes through the video counting station before connecting into the surface collector bypass pipe at the juvenile facility.

Evaluation methods

Radiotelemetry

The radio telemetry study was conducted by Bioanalysts, Inc. (formerly Don Chapman Consultants) of Redmond, Washington and LGL Limited of Canada in 1996 and 1997. Fish migration into the forebay was monitored with aerial antennas along the powerhouse, forebay wall, spillway, and west bank. Movement near the surface collector entrance was monitored with both aerial and underwater antennas. Monitoring began in mid-April and ended the first week in June each year. Approximately 30–50 juvenile chinook and steelhead were tagged with Lotek MCFT-3K and MCFT-3G digital radio transmitters every other day starting in mid-April. After tagging, fish were allowed to recover over a 24 h period and released about 3 km upstream of the dam. Stevenson et al. (1997) discuss telemetry equipment, data analysis, and results in more detail.

Passive integrated transponder tag study

Tagging began in mid-April and ended in mid-July. After tagging, fish were allowed to recover over a 24 h period and released in the forebay, adjacent to Unit 11, approximately two-thirds of the way across the forebay from the powerhouse. Four 0.61 m PIT tag detectors were installed in the bypass pipe (immediately upstream of the video monitoring station) in 1996. We installed two 0.30 m PIT tag detectors in the gatewell collector system bypass pipe, parallel to the 0.61 m detectors in 1997. In 1996, fish recovered in the headgate or surface collector were released at the alternative release site (ARS) which consisted of a 5,753 L holding tank that was plumbed into the main transport pipe. The ARS system had two 0.15 m PIT tag detectors installed in it. The 0.15 m detectors were used, because the 0.61 m detectors were not functioning in 1996.

Marked fish releases

To assess whether passage through the surface and gatewell collectors affected fish condition, we marked fish and released them back into the systems. All marked fish were examined before release for descaling and injury; only fish showing no descaling or injuries were marked and subsequently released. Upon recapture, fish were again examined to determine whether passage through the collectors caused descaling and injury.

Headgate salvage operation (1995 and 1996)

Using a crowder and an extractor box, we captured fish in each of the three headgate slots of turbine Unit 1. Headgate slots were dipped through mid-August to prevent fish from being injured or stressed due to excessive time spent in the slots. Fish were transported and released off a barge in the tailrace (1995) or into the ARS (1996), which was plumbed into the main juvenile bypass pipe. We added a weir and transport system in 1997 that made dipping unnecessary.

Results

Passage efficiency of surface collector

We tested fish passage efficiency of the surface collector in 1996 and 1997 by releasing fish with PIT tags and recovering fish in the surface and gatewell collector systems. In 1996, we released 1,481 juvenile chinook and 1,491 juvenile steelhead, recapturing 331 chinook (22.3%) and 406 steelhead (27.2%; Table 1; Figure 7). We recaptured most of the chinook in the surface collector, but most of the steelhead were captured in the headgate slot (Figure 8). We increased our capture efficiency in 1997: 35.1% for chinook (n = 1,298) and 56.7% for steelhead (n = 1,478; Figure 7). We also released juvenile sockeye salmon *O. nerka* in 1997 and recaptured 14.7% (n = 789; Table 1). In contrast to 1996, we recaptured most of the chinook and steelhead in the surface collector (Table 1; Figure 8).

Table 1. Summary of PIT tag releases and recaptures of juvenile salmonids at Rocky Reach Dam, 1996 and 1997 (Peven et al. 1996).

			Recaptures			
Year	Released species	Number	Surface collector	Diversion screens	Combined	Percent total
1996	Chinook (CH)	1,481	183	148	331	22.3
1997	Chinook	1,298	293	163	456	35.1
1996	Steelhead (STL)	1,491	110	296	406	27.2
1997	Steelhead	1,478	685	153	838	56.7
1997	Sockeye	789	68	48	116	14.7
1996	CH + STLD	2,972	293	444	737	24.8
1997	CH + STLD	2,776	978	316	1,294	46.6

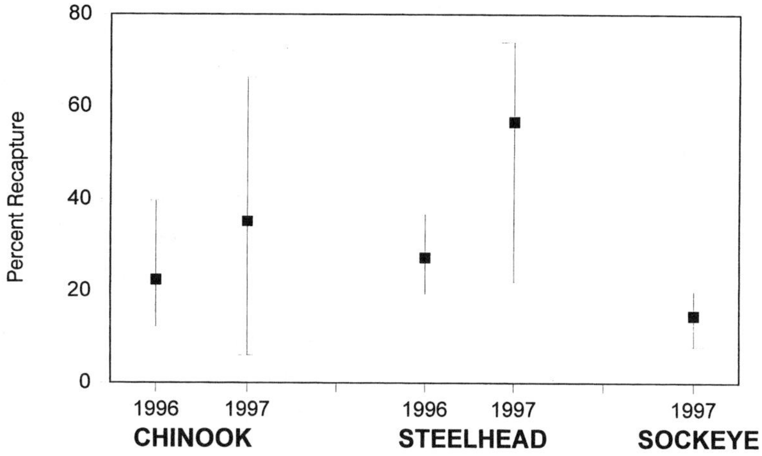

Figure 7. The mean (with individual test ranges) recapture rate of PIT tagged juvenile salmonids at Rocky reach Dam, 1996 and 1997 (Peven et al. 1996).

Based on PIT-tag recoveries, the total passage efficiency of the fish bypass system (diversion screens and surface collector) for steelhead and chinook increased by almost 22% from 1996 to 1997 (Table 1). We attribute this increase to the change of hydraulics both within and immediately upstream of the collector entrance.

Fish movement and behavior

In 1996, we inserted radio tags into 202 juvenile chinook and 193 juvenile steelhead. Based on our monitoring, most of the fish entered the forebay from the west shoreline area and moved across the forebay toward the powerhouse (Figure 2). Once at the powerhouse, most of the fish moved parallel to the powerhouse and downstream towards Unit 1 near the vi-

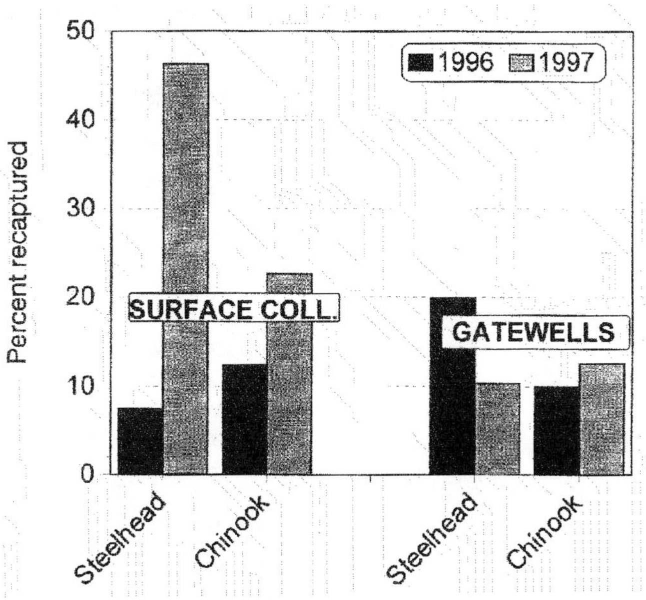

Figure 8. Recapture rates of radio-tagged juvenile chinook and steelhead from the Rocky Reach fish bypass system (Peven et al. 1996).

cinity of the surface collector entrance. Fish detected immediately upstream of the surface collector entrance did not readily move into and through the collector. In fact, most of the fish that entered the collector later exited and passed the dam via another route (Stevenson et al. 1997; see below). Some of the fish exhibited "milling" behavior, as they failed to exhibit directional movement.

Although in 1996 the median forebay residence time (elapsed time between first and last detection at the dam) was low for juvenile chinook and steelhead (0.14 and 3.28 h, respectively), some fish spent days in the forebay before passing the dam (414 and 176 h for one chinook and one steelhead, respectively; Table 2). In 1997, the median residence time was much shorter (0.04 and 0.25 h for juvenile chinook and steelhead, respectively), although some fish still took much longer (Table 2).

In 1997, we tagged 211 steelhead and 186 chinook. In contrast to 1996, most of the fish approached the dam either near the spillway or the upstream turbine units (J. Stevenson, BioAnalysts, personal communication). The difference in fish behavior between 1996 and 1997 was probably because large volumes of water passed through the spillway and upstream turbine units in 1997, when river flows exceeded powerhouse capacity. Once the fish reached the project, most of them exhibited the same behavior as in 1996 (moving down the powerhouse towards the cul-de-sac), although more fish passed through the upper units in 1997 compared to 1996 (Stevenson, personal communication).

The hydraulic changes within the collector between 1996 and 1997 appeared to improve the effectiveness of the collector. The percentage of juvenile steelhead that passed all the way through after entering the collector increased from 28% to 73% (Stevenson, personal communication).

We also tested to see if a narrower, higher-velocity entrance was more efficient in entraining fish than a slower, wider entrance. We found that there was no difference between the two configurations in the proportion of fish observed in front of the collector that passed into the entrance. Contrary to what was anticipated, slightly fewer fish exited the collector once they entered it with the narrower, higher velocity configuration.

Marked fish releases

Over the three year period when fish were examined, marked, and released, descaling and injury has generally run less than 5.0%.

Table 2. Detection time of radio-tagged juvenile chinook and steelhead within the Rocky Reach Dam telemetry system in hours (elapsed time between first and last detection at the project), 1996 and 1997 (Stevenson et al. 1997; J. Stevenson, BioAnalysts, personal communication).

Species	Year	Number detected	Mean	Median	Range	Standard deviation
Chinook	1996	122	8.22	0.14	0.01–414.3	39.75
Steelhead	1996	169	11.71	3.28	0.01–176.3	27.63
Chinook	1997	88	0.77	0.04	0.00–21.5	2.68
Steelhead	1997	162	8.98	0.25	0.00–292.3	33.29

Discussion

The biological evaluation of the surface collector at Rocky Reach in 1995 was limited to the enumeration of total passage, subsamples of species composition, injury rate, and behavior. The main focus was a "proof of concept" evaluation; thus we did not evaluate collection efficiency. Having concluded the "proof of concept" in 1995, the focus of the biological evaluation in 1996 and 1997 was to determine collection efficiency, and to evaluate the movement of fish in the forebay and near the surface collector entrance and relate that information to collection efficiency.

Our observations in 1996 suggested that the fish bypass system, especially the surface collector, was less effective than in 1995 (Peven et al. 1996). We used hydraulic and mathematical models to help us determine what the major differences were between 1995 and 1996. The modeling showed that in 1995, the water immediately upstream of the surface collector entrance was flowing directly into it; while in 1996, flow was skewed across the face of the entrance. Based on these findings, we removed the floor and sloping wall to go back to the 1995 configuration. Incremental model testing showed the best alternative (to achieve flow conditions like those that appeared in 1995) was to totally remove the floor and sloping wall.

The hydraulic changes caused by the floor and sloping wall appeared to be associated with differences in fish movement patterns. In 1996, a split-beam hydroacoustic transducer scanned the forebay on a barge upstream of the surface collector entrance. Data from the hydroacoustic monitoring suggested that fewer fish used the surface collector in 1996 than in 1995 (Steig and Adeniyi 1996). The relative difference between both years was based on the horizontal distribution of fish across the dam. In 1994, the first year we deployed the barge in the forebay for the split-beam system, large numbers of fish were observed near the forebay wall area (in the vicinity between the surface collector and the exit of the adult fishway). In 1995, when we installed the surface collector (without the floor and sloping wall), we did not observe large numbers of fish near the forebay wall. The data collected in 1996 (Steig and Adeniyi 1996) once again showed large numbers of fish downstream (along the forebay wall) of the surface collector entrance. Also in 1995, the direction of fish, as determined by the split-beam transducer, was toward the surface collector entrance, while in 1996 it was more toward the west, crossing the entrance. These data suggest that the 1995 hydraulic conditions were more conducive to guiding fish than the 1996 configuration. After modifying the system back to 1995 hydraulic conditions, we again saw no sign of fish near the forebay wall in 1997 (R. Adeniyi, Hydroacoustic Technology, Inc., personal communication), suggesting that the fish were spending less time in the forebay and readily passing through the surface collector; this conclusion is also supported by the decrease in the median forebay residence time of radio-tagged smolts in the forebay in 1997.

The results from the radio telemetry study in 1996 showed that a large number of fish entered the surface collector entrance but returned to the forebay instead of passing through the system. Because of this observation, we reevaluated the flow characteristics inside the entrance. We determined that velocity was decelerating as the water flowed further inside the entrance (Figure 9). Since a decelerating velocity is counter to

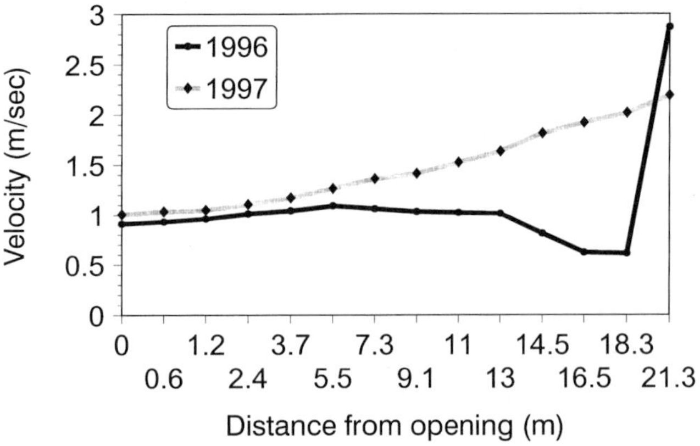

Figure 9. Velocity profile of the surface collector entrance in 1996 and 1997 (C. Sweeney, ENSR Engineering, personal communication).

what we think will increase fish passage, we made major modifications to the entrance of the surface collector to increase velocity in 1997 and were successful (Figure 9). We believe the increase in steelhead passage through the entrance in 1997 resulted from these hydraulic changes.

Our primary purpose of extending the floor in 1996 was to see if we could induce the fish that were still going into Unit 1 to remain in the surface waters, and increase the likelihood of guidance into the collector. We have not been able to determine whether the floor kept fish nearer the surface, because we were not able to deploy underwater radio antennas in arrays near the zone of influence of the extended floor. This concept may be feasible in some locations, but the overall effect at Rocky Reach was detrimental to fish guidance because it changed the approach hydraulic conditions to the entrance of the surface collector.

We tested the concept of a higher velocity entrance in 1997 by installing a removable steel structure in front of the opening for two separate weeks during the six week study. We anticipated that because of flow separations created by its presence, that fish would be less likely to pass all the way through the entrance. In fact, we found that slightly more fish passed all the way through the system when the slot narrowing device was present. However, we found no difference in the percentage of fish detected immediately upstream of the entrance that entered the collector. This suggests that a higher velocity entrance may be worth exploring. One of our research goals at Rocky Reach in 1998 is to test an entrance condition that has a velocity of about 0.9 m^3/s at the opening and increases rapidly (within 6 m) to >2.1 m^3/s, while avoiding sudden flow acceleration, which may cause an avoidance response.

Johnson et al. (1997) proposed guiding premises for designing surface collectors. These are

- fish follow the bulk of the flow as they approach a particular dam,
- salmonid smolts prefer not to sound to follow a passage route,
- smolts have the ability to discover an entrance, and
- the entrance conditions do not dissuade the fish from either entering or passing through (Johnson et al. 1997).

In relation to these premises, forebay surface currents at Rocky Reach are directed towards the cul-de-sac area, defined by the juncture of the powerhouse and forebay wall (Figure 2). The water entering the upstream units (8–11) and surface currents run parallel to the powerhouse (ENSR Consulting and Engineering 1995). We believe this condition leads to fish concentrating in the cul-de-sac area. If fish are concentrated in one section of the forebay at a dam, their opportunity to discover the entrance of a surface collector is probably greater, provided that both the approach and entrance hydraulic conditions are favorable (a lesson we learned in 1996). This may be why the results of fish passage efficiency of the fish bypass system at Rocky Reach Dam have appeared more promising than at other locations on the Columbia River (Kumagai et al. 1996; Johnson et al. 1997). The success of Wells Dam can be contributed, in part, to its unique design (spillways on top of, and not separate from turbines), which basically gives the fish no other options for fish passage (horizontally), while the entrance conditions favor fish entrainment and passage (Johnson et al. 1997)

With further improvement in fish passage efficiencies, and construction of additional entrances, this hybrid fish passage system (surface collector and diversion screens) appears to be the preferred method for juvenile salmonid bypass at Rocky Reach Dam. Future entrances of the surface collector will be placed in locations in the forebay that will enhance the opportunity of discovery. Entrance location, coupled with attractive entrance hydraulic conditions, both immediately upstream and within the entrance, should ensure efficient fish passage. By bypassing a large percentage of the juvenile salmonids safely around the dam, in conjunction with predator abatement in the tailrace, Chelan will increase the survival rate of migrating fish passing Rocky Reach Dam.

Acknowledgments

We thank the numerous people that have spent countless hours working on this project over the years. Some of the more prominent include, D. Nason, S. Hays, B. Christman, B. Bickford, K. Truscott, L. Lorrain (and crew), and D. Beardsley of Chelan County Public Utility District. D. Hay, D. Weitkamp, C. Sweeney, B. Gatton, D. Whitney, and M. Erho all have provided wise and useful council over the years. Steve Hays reviewed a draft of this paper.

References

ENSR Consulting and Engineering. 1995. Hydraulic model studies of downstream juvenile fish bypass systems at Rocky Reach Dam. Forebay field data collection, model preparation, and investigation of diversion screen system for 1994 prototype tests. Final Report of ENSR to Chelan County Public Utility District, Wenatchee, Washington.

Johnson, G. E., A. E. Giorgi, and M. W. Erho. 1997. Critical assessment of surface flow bypass development in the lower Columbia and Snake rivers. Completion Report to the U.S. Army Corps of Engineers, Portland and Walla Walla Districts.

Kudera, E. A., and C. M. Sullivan. 1993. Evaluation of the smolt bypass system at Wells Dam in 1992. Final Report of BioSonics, Inc. for Douglas County Public Utility District, East Wenatchee, Washington.

Kumagai, K. K., B. H. Ransom, and H. A. Sloan. 1996. Effectiveness of a prototype surface flow attraction channel for passing juvenile salmon and steelhead trout at Wanapum Dam during summer 1996. Final Report of Hydroacoustic Technology, Inc. to Grant County Public Utility District, Ephrata, Washington.

Peven, C. M., and B. G. Keesee. 1992. Rocky Reach fish guidance system 1992 developmental testing. Chelan County Public Utility District, Final Report, Wenatchee, Washington.

Peven, C. M., and A. M. Abbott. 1994. Rocky Reach fish guidance system 1994 developmental testing. Chelan County Public Utility District, Final Report, Wenatchee, Washington.

Peven, C. M., A. M. Abbott, and B. M. Bickford. 1995. Biological evaluation of the Rocky Reach surface collector 1996. Chelan County Public Utility District, Final Report, Wenatchee, Washington.

Peven, C. M., T. R. Mosey, and K. B. Truscott. 1996. Biological evaluation of the Rocky Reach surface collector 1996. Chelan County Public Utility District, Final Report, Wenatchee, Washington.

Raemhild, G., T. Steig, R. Riley, and S. Johnston. 1984. Hydroacoustic assessment of downstream migrating salmon and steelhead at Rocky Reach Dam in 1983. Final Report of Biosonics, Inc. to Chelan County Public Utility District, Wenatchee, Washington.

Steig, T. W., and A. E. Sullivan. 1991. Hydroacoustic evaluation of the horizontal distribution of juvenile salmon and steelhead across the powerhouse at Rocky Reach Dam during 1990. Final Report of Hydroacoustic Technology, Inc. to Chelan County Public Utility District, Wenatchee, Washington.

Steig, T. W., R. Adeniyi, and T. K. Iverson. 1995. Hydroacoustic evaluation of the behavior of juvenile salmon and steelhead approaching the powerhouse in the forebay of Rocky Reach Dam during 1994. Final Report of Hydroacoustic Technology, Inc. to Chelan County Public Utility District, Wenatchee, Washington.

Steig, T. W., and R. Adeniyi. 1996. Hydroacoustic evaluation of the fish passage through units 1-11, spillways 3-5, and the juvenile surface collector at Rocky Reach Dam in the spring and summer of 1996. Final Report of Hydroacoustic Technology, Inc. to Chelan County Public Utility District, Wenatchee, Washington.

Stevenson, J. R., A. E. Giorgi, W. R. Koski, K. K. English, and C. A. Grant. 1997. Evaluation of juvenile spring chinook and steelhead migratory patterns at Rocky Reach and Rock Island Dams using radio telemetry techniques, 1996. Final report of BioAnalysts, Inc., LGL Limited, and Grant Systems Engineering to Chelan County Public Utility District, Wenatchee, Washington.

Tracey W. Steig and Rowland Adeniyi

Hydroacoustic Evaluation of Fish Passage through a Prototype Surface Bypass Collector at Rocky Reach Dam

6

Salmon and steelhead runs on the Columbia River and its tributaries have been in decline due to several factors, including the operation of hydroelectric dams. In 1995, a prototype surface collector (PSC) was constructed at Rocky Reach Dam to bypass fish around the turbine units. The PSC uses considerably less flow than a turbine unit and passes fish with minimal stress. Juvenile salmonids were monitored using hydroacoustic methods, which permitted efficient, nonobtrusive sampling of the fish. Sampling was conducted at the site in 1995 and 1996 to estimate the proportion of juvenile fish detected in front of the PSC using a combination of single-beam and split-beam hydroacoustic techniques. For the two years of operation, the proportion of fish traveling toward the PSC entrance as a function of depth varied from 76% in 1995 to 70% in 1996. The horizontal distribution of fish passage across the dam showed that most fish were detected in the vicinity of the PSC for both study years. The vertical distribution of the fish detected in front of the PSC was normally distributed throughout the water column. In summary, hydroacoustic studies in 1995 and 1996 showed that the PSC provided efficient bypass route for downstream migrating juvenile salmonids at Rocky Reach Dam.

Introduction

Due to several factors, wild runs of pacific salmon and steelhead trout *Oncorhynchus* spp. on the Columbia River and its tributaries have been declining in magnitude. A factor in this decline has been the construction and operation of mainstem hydroelectric dams. While most downstream migrating juvenile salmon pass safely through a single dam, the cumulative mortality that results from passing through several dams can be substantial (Schwiebert 1977).

For several decades, the development of efficient techniques to bypass downstream migrant salmon and steelhead around hydroelectric turbines has been a high priority in the Columbia River Basin, including Rocky Reach Dam. Diverting downstream migrant salmon and steelhead around turbine intakes may increase their survival rate, and help restore the region's anadromous fish populations.

From 1985 to 1994, Chelan County Public Utility District No. 1 (Chelan PUD) tested a variety of prototype rotating and passive diversion screens in an attempt to increase survival of downstream migrants passing Rocky Reach Dam. These prototypes guided less than 50% of the fish entering the turbine intakes (Truscott and Hays 1988, 1989; Peven et al. 1991). As an alternative, Chelan PUD constructed a prototype surface collection (PSC) system based on the concept of the Wells Dam spillway baffles, a successful bypass structure 60.7 km upstream of Rocky Reach Dam. This prototype was constructed and installed in the forebay area near the powerhouse in front of Turbine Unit 1 in the fall of 1994. The PSC was operated for the first time during the 1995 spring and summer out-migration of juvenile salmonids.

Before the 1996 out-migration, the PSC was modified to include a flat platform attached to the bottom of the PSC at a depth of approximately 17.7 m that extended 18.3 m out from the powerhouse wall up to the north end of Turbine Unit 2. In addition, sloping walls were installed on the PSC. The PSC entrance remained the same during the 1995 and 1996 out-migration. An additional adjustable venturi gate was added in 1996 to the two existing in 1995. This increased the water velocity through the PSC entrance from 59.5 m^3/s to a maximum of 66.5 m^3/s in 1996.

In 1995 and 1996, hydroacoustic studies were conducted to monitor fish passage into the PSC, turbine units, and spill gates at Rocky Reach Dam during the spring and summer out-migration (Ransom 1991; Ransom and Steig 1994).

The primary objective of the studies was to evaluate the PSC's ability to attract downstream migrants away from the turbine intakes and into the bypass system. Specific objectives of the hydroacoustic study were to estimate:

- the horizontal and vertical distribution of fish passage,
- the hourly fish passage and
- the direction of fish movement in front of the PSC.

Study Site

Rocky Reach Dam is on the Columbia River 11.3 km north of Wenatchee, Washington at river kilometer 764.3 (Figure 1). The dam's spillway is perpendicular and its powerhouse parallel to river flow (Figure 2). The powerhouse is 331.6 m long and contains 11 vertical Kaplan turbines, numbered from south to north. Turbine Units 1–7 have a rated generating capacity of 116 MW each.

Turbine Units 8–11 are rated at 125 MW each. Each turbine has three rectangular intakes, 6.1 m wide by 15.2 m high at the headgate slot. The spillway is over 228.6 m long and has 12 automatic spill gates. Each gate is 15.2 m wide and approximately 18.3 m deep to the spill gate ogee.

Methods

Hydroacoustic equipment and operation

Two different hydroacoustic systems were used at Rocky Reach Dam in 1995 and 1996. The single-beam system consisted of two 420 kHz echo sounder/transceivers, two multiplexer/equalizers, three chart recorders, two oscilloscopes, and 6° circular and 6° × 12° elliptical transducers.

A split-beam hydroacoustic system was also used at Rocky Reach Dam to monitor fish movement approaching the PSC. The system consisted of a 200 kHz split-beam echo sounder and a 200 kHz circular 6° split-beam transducer. A more detailed description of the split-beam theory and technique is discussed by Steig et al. (1994, 1995).

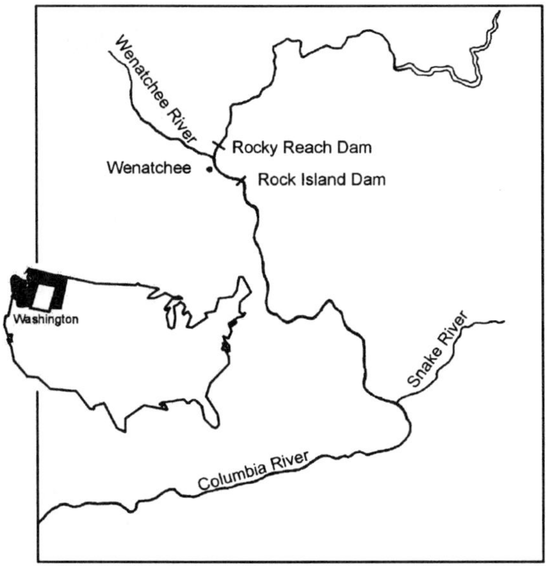

Figure 1. Location of Rocky Reach Dam on the Columbia River in Washington State.

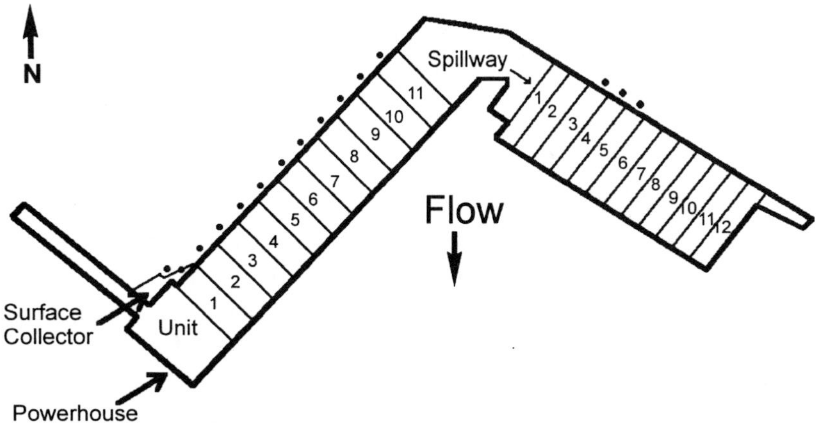

Figure 2. Plan view of Rocky Reach Dam showing transducer mounting locations.

Both systems were calibrated and equalized so that only targets of −56 dB or larger would be detected. Thus, targets of −50 dB or larger would be detected out to the −3 dB points (one-way propagation) of the transducer beam pattern. This threshold was sufficient to detect the smallest juvenile salmonids present in the river.

Study design and dam operations

The PSC and Turbine Units 1–5 were monitored throughout the spring and summer study periods in 1995. The PSC and Turbine Units 1–11, and Spill Gates 3–5 were monitored throughout the spring and summer study periods in 1996. The operation of the turbine units was dependent on power demand and the river flow at any given time. However, Turbine Units 1–5 had the highest priority of operation.

Transducer deployment

Circular and elliptical beam transducers were placed immediately in front of the PSC, turbine units, and spillway. Each transducer was sampled approximately 10 min/h in 1995 and 4 min/h in 1996. The powerhouse transducers were mounted at a water depth of approximately 34.4 m, 0.9 m in front of the intake pier noses (Figure 3). All the powerhouse transducers were aimed upward at approximately 5° upstream. The PSC transducer was mounted at a water depth of 17.7 m, aimed vertically upward. The spill gate transducers were mounted at a water depth of 3.1 m, 1.5 m in front of the spill gate pier noses and centered in the spill gate. These transducers were aimed downward at approximately 15° upstream.

Data collection and analysis

Hydroacoustic data were collected in 1995 from 17 April through 27 May for the spring study and 23 June through 7 July for the summer study. In 1996, hydroacoustic data were collected from 22 April through 1 June for the spring study and 24 June through 13 July for the summer study. The data were collected 5 d/week, 24 h/d beginning each Monday morning at 0700 hours and ending on Saturday morning at 0700 hours.

The horizontal distribution of fish passage was calculated for the PSC, turbine units, and spill gates as the percentage of fish or flow for the entire dam that passed through an intake. Percentage distributions were first calculated on a daily basis and then summarized by the sampling periods to remove the effect of daily variation in passage. Operations of the PSC, turbine units, and spill gates were calculated for each intake as the percentage of hours per day that each system was operating, out of 24 h of possible operations. Percentage operations for each intake are presented on a daily basis and averaged across days for the entire sampling periods.

Hourly estimates of fish passage were averaged over each study period and presented as the average percentage fish passage for each of the 24 h of the day.

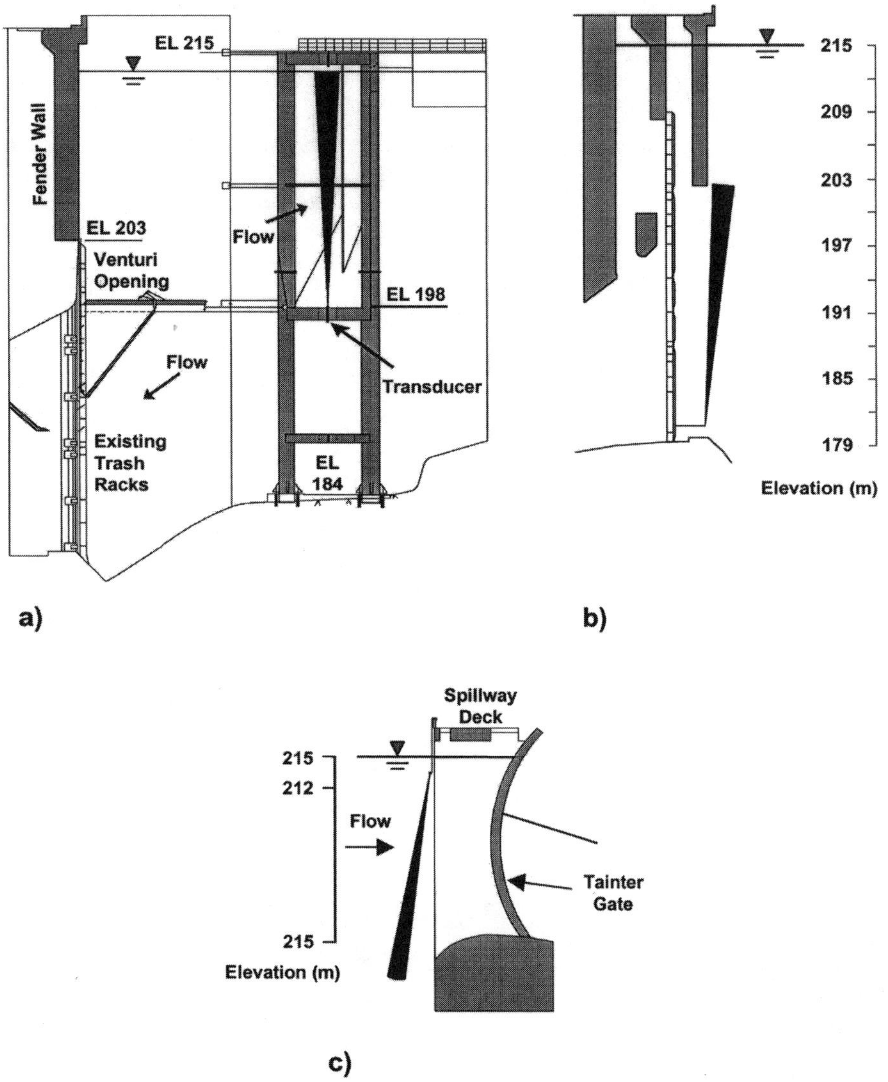

Figure 3. Cross-sections showing the transducer mounting locations for a) surface collector, b) turbine units, and c) spill gates.

Vertical distributions were calculated from the average range at which a fish passed through the acoustic beam. Fish distributions, as a function of range from the transducer, were estimated as the percentage of fish passing through the acoustic beam for each 1 m stratum.

Single-beam hydroacoustic techniques were used to estimate fish entrainment at the PSC, turbine units, and spill gates. However these techniques were unable to determine fish movement direction. Fish detected directly in front of the turbine units were assumed to be passing into the intakes, however, there were indications (Steig and Adeniyi 1995) of fish milling in the area directly upstream of the PSC. Therefore, the proportion of fish reported for the PSC should be considered as fish that were available for collection as opposed to fish that were guided by the PSC. To complement the single-beam techniques, split-beam hydroacoustic techniques were used to determine the directionality of the fish in the area directly upstream of the PSC (Steig and Johnston 1996; Steig and Iverson 1998).

Results

Single-beam monitoring

The horizontal distribution results showed the highest proportion of fish were detected at the PSC in 1995 with 48.0% of the fish and 3.9% of the flow (Figure 4; Table 1). The lowest fish passage was through Turbine Unit 5 in 1995 with 2.1% fish passage and 16.3% flow. Overall fish passage and flow were higher through turbines (52.0% fish passage and 98.1% flow) than through PSC (48.0% fish and 1.9% flow). During this period, the turbine units and the PSC were operated 57.7% and 73.1% of the time, respectively.

The horizontal distribution of fish passage through the PSC, turbine units, and spill gates was similar for both years. In 1996, the highest proportion of fish was detected at the PSC with 26.8% of the fish and 1.1% of the flow. The lowest fish passage was through Turbine Unit 11 in 1996 with 0.2% fish passage and 8.1% flow. Overall fish passage and flow were higher through turbines (69.8% fish passage and 81.7% flow) than through spill bays (3.4% fish passage and 17.2% flow). The average turbine units and PSC operations were 86.6% and 98.9%, respectively.

Spillway fish passage was not monitored in 1995. In 1996, spillway fish passage averaged 3.4% for a spill gate operation of 61.6%. Spillway fish passage in 1996 was not limited to the monitored spill gates (Spill Gates 3–5), because of the high river flows during the spring study. During the spring study period, spill was limited to Spill Gates 3–5 only from 9–16 May (Steig and Adeniyi 1996).

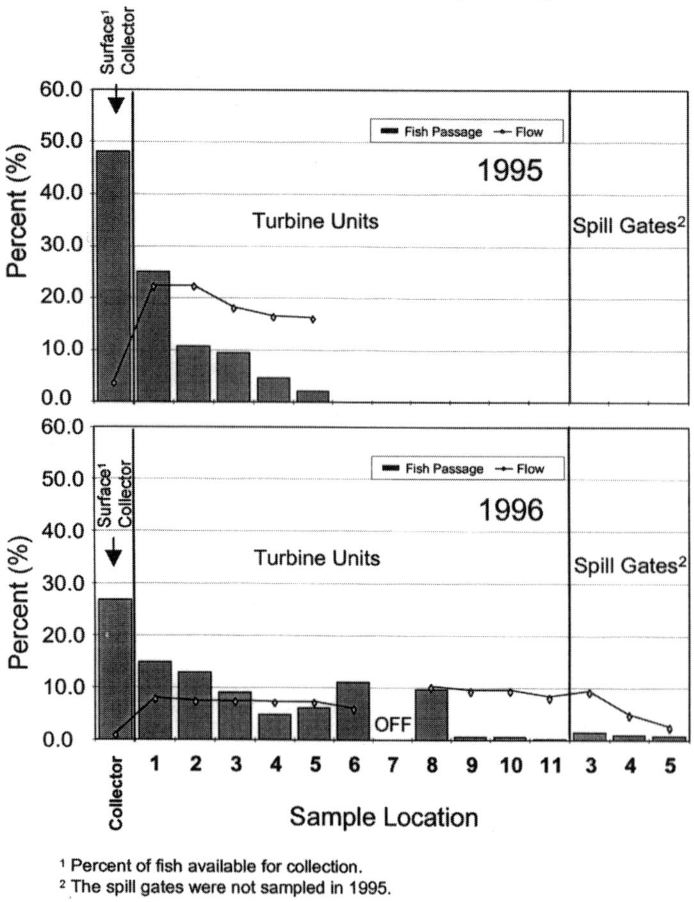

Figure 4. Horizontal distribution of fish passage estimates for the surface collector, turbine units, and spill gates.

Table 1. Percentage of fish passage and turbine operations at Rocky Reach Dam during the 1995 and 1996 studies.

Description	1995			1996		
	Surface collector [a]	Turbine units	Spill gates [b]	Surface collector	Turbine units	Spill gates
Fish passage	48.0	52.0		26.8	69.8	3.4
Flow	1.9	98.1		1.1	81.7	17.2
Operations	73.1	57.7		98.9	86.6	61.6

[a] Percentage of fish available for collection.
[b] The spill gates were not monitored in 1995.

The summer horizontal distribution showed similar results to the previously studies. The results showed Unit 1 passed more fish than Units 2 and 3. These results suggest that the surface collector had fewer fish detected near the opening during the summer (29.9%) as compared to the spring (66.1%).

Fish passage was higher at night during the sampling periods. The highest proportion of fish at all locations occurred between 2100 and 2300 hours in 1995 (Figure 5). During 1995, 59.3% of fish detection at the PSC occurred during the daytime. In 1996, approximately 45.1% of the fish passed through the turbines during the daytime.

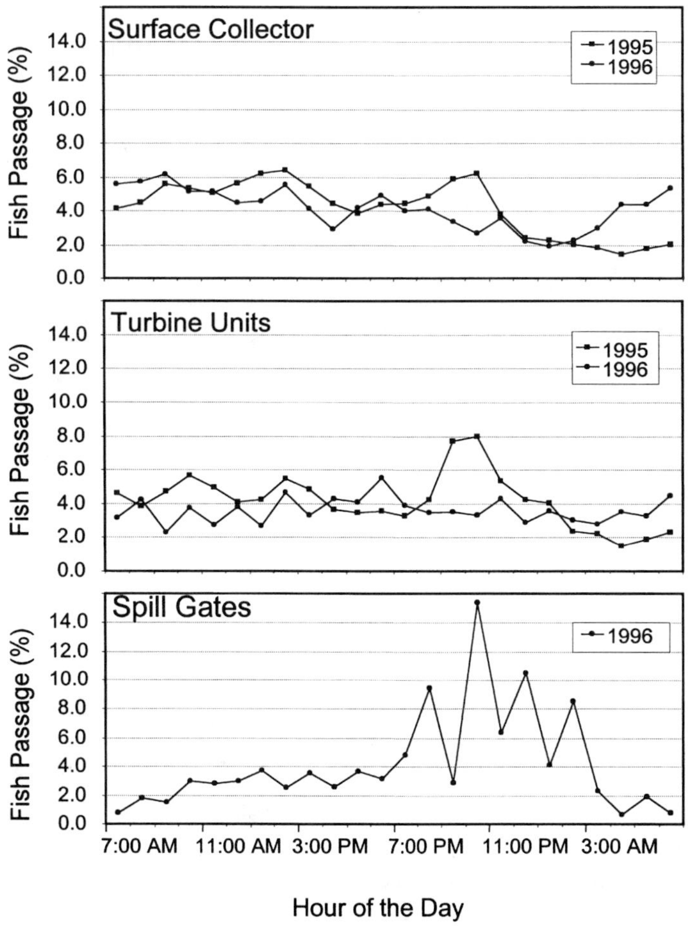

Figure 5. Hourly percent fish passage for the surface collector, turbine units, and spill gates.

The highest proportion of fish detected at the PSC in 1996 occurred between 0800 and 0900 hours and the lowest proportion occurred between 1200 and 0200 hours. The result suggests that the fish were primarily surface oriented during the daytime and were detected at the PSC. During the nighttime, the fish tended to be distributed deeper in the water column and detected entering the turbine units and the spill gates.

The vertical distribution of fish in front of the PSC was normally distributed throughout the depth of the entrance except for a spike at elevation of 207 m for the 1996 study (Figure 6). The vertical distributions of the fish entering the turbine units showed the majority of the fish were entering the turbines in the top 4.0 m of the intakes. From elevation 190 m to 200 m, the vertical distribution of the fish was very consistent entering the turbine units. The vertical distribution of fish was skewed higher in the water column with the majority of the fish entering the spill gates in the top 7.6 m. From elevation 194 m to 206 m, the vertical distributions of fish entering the spill gates were relatively evenly distributed.

Split-beam surface collector monitoring

In 1995, more fish (76%) were moving toward the entrance of the PSC (Figure 7) than away (23%). Overall, the proportion of fish traveling toward the PSC entrance increased from approximately 70% during the spring to approximately 90% during the summer. In 1996, similar to 1995, the majority of the fish (70%) were moving toward the entrance of the PSC throughout the water column. Also in that year, 28% of the fish were moving away from the PSC entrance, compared to 23% for 1995. There were 2% of the fish in 1996 and 1% of the fish in 1995 in which the direction of movement could not be determined.

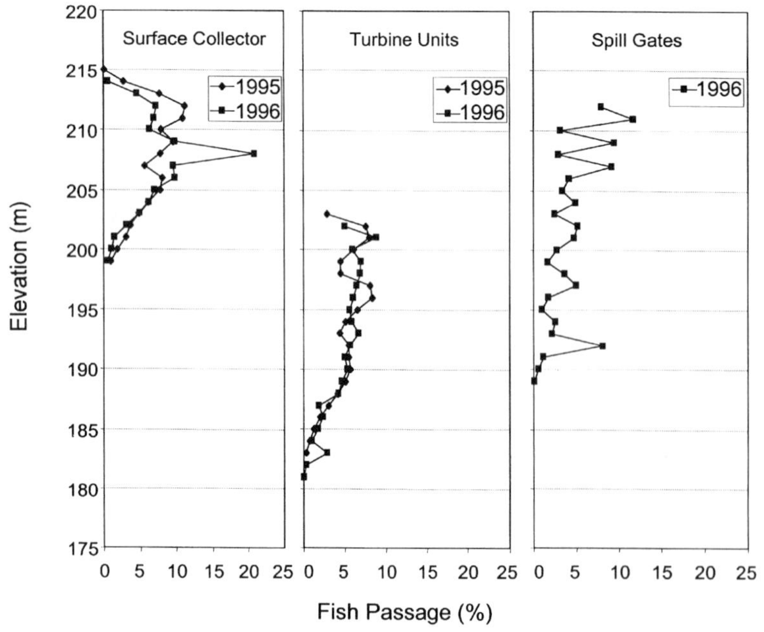

Figure 6. Vertical distribution of fish passage for the surface collector, turbine units, and spill gates.

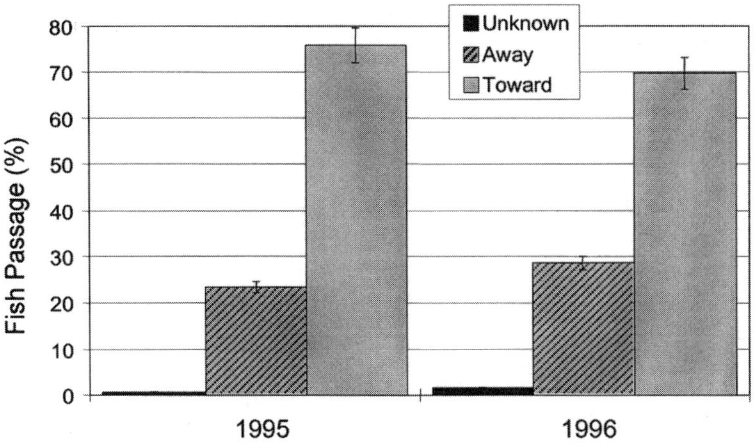

Figure 7. Proportion of fish detected directly upstream of the surface collector moving toward or away from the entrance. Error bars are 1 SE.

Discussion

Single-beam monitoring

Horizontal distribution across the dam in 1995 showed the proportion of fish passage shifted from Turbine Units 1 and 2 to the PSC. In contrast, smaller proportions of fish were detected in front of the PSC in summer 1995. Studies at Rocky Reach Dam and The Dalles Dam have shown deeper vertical distribution of fish during the summer as compared to spring (Steig and Johnson 1986; Steig and Adeniyi 1995). This could explain the decrease in effectiveness of the PSC during the summer.

During the spring in 1996, fish passage at Turbine Units 6 and 8 were unusually high compared to previous years (Steig et al. 1994, 1995). The high spring fish passage at Turbine Unit 8 could be related to the installation of a new trashboom in the powerhouse forebay. During construction, numerous floating pontoon sections were moored to the face of the dam upstream of the north end turbine units (Turbine Units 8–11). These pontoons may have influenced fish passage through Turbine Unit 8.

Another explanation for the high fish passage through Turbine Units 6 and 8 could be due to Turbine Unit 7 not operating. Previous studies have shown that fish passage increased in the turbine units located next to turbine units not operating. In 1983, when Turbine Unit 7 was also shut down, a very similar horizontal distribution was seen, with increased fish passage through Turbine Units 6 and 8, as compared to the other turbine units

The 1996 vertical distribution of fish in front of the PSC was normally distributed throughout the depth of the entrance with a slight increase in the middle of the PSC. One explanation for the normal distribution may be attributed to the installation of a flat platform attached to the bottom of the PSC. Another explanation may be due to the installation of sloping walls around the PSC and the construction of an additional adjustable venturi gate below the PSC. All these modification increased water velocity throughout the PSC entrance hence distributing the fish evenly throughout the PSC.

Split-beam surface collector monitoring

The proportion of fish traveling toward the PSC entrance in 1995 increased from approximately 70% during the spring to approximately 90% during the summer. Similar results between the spring and summer studies were also reported for the biological evaluation of the PSC (Peven et al. 1995). These differences may be an indication of the differences in the fish behavior and species composition between the spring and summer migration periods.

In 1996, the proportion of fish traveling toward the PSC entrance was similar (70%) during the spring and summer studies. The top and bottom of the PSC detected the smallest proportion of fish moving away from the PSC. Conversely, the middle of the water column (elevation 206–209 m) in front of the PSC was the area where higher proportions of fish were moving away from the PSC. This may be due to a structural cross member at the entrance of the PSC at the 207 m elevation.

In summary, hydroacoustic studies in 1995 and 1996 showed that the PSC provided efficient bypass route for downstream migrating juvenile salmonids at Rocky Reach Dam. The majority of monitored outmigrants passed through the PSC utilizing a minimum of project outflow. More juvenile salmonids used the bypass system in 1995 as compared to 1996. Hydraulic conditions upstream of the PSC entrance may have decreased its effectiveness of the PSC in 1996. Hydraulic and mathematical models will be used to determine the major differences in hydraulic conditions between 1995 and 1996. Hydroacoustic studies will continue in 1997 and 1998 to monitor PSC over time and varying conditions.

Acknowledgments

The study was financed in part by the Chelan Public Utility District (PUD). This publication does not necessarily reflect the views and policies of the Chelan PUD. The following Chelan PUD personnel were instrumental in obtaining the materials and facilities needed: Richard Nason, Steve Hays, Chuck Peven, Brett Bickford, and the Central Maintenance mechanics crew.

References

Peven, C. M., S. G. Hays, and K. B. Truscott. 1991. Rocky Reach prototype fish guidance system 1990 developmental testing. Final Report to Chelan County Public Utility District, Wenatchee, Washington.

Peven, C. M., A. M. Abbott, and B. M. Bickford. 1995. Biological evaluation of the Rocky Reach surface collector 1995. Chelan County Public Utility District, Wenatchee, Washington.

Schwiebert, E., editor. 1977. Columbia River salmon and steelhead. American Fisheries Society, Special Publication 10, Bethesda, Maryland.

Raemhild, G. A., T. W. Steig, and R. Riley. 1983. Hydroacoustic assessment of downstream migrating salmon and steelhead at Rocky Reach Dam in 1982. Report by BioSonics, Inc. to Chelan County Public Utility District, Wenatchee, Washington.

Ransom, B. H. 1991. Using sound waves to monitor fish entrainment. HydroReview 10(4):104-115.

Ransom, B. H., and T. W. Steig. 1994. Using hydroacoustics to monitor fish at hydropower dams. Lake & Reservoir Management 9(1):163-169.

Steig, T. W., P. A. Nealson, and M. A. Nason. 1994. Hydroacoustic evaluation of juvenile salmon and steelhead approaching turbine units 1-11 at Rocky Reach Dam during the summer of 1993. Report by Hydroacoustic Technology, Inc. to Chelan County Public Utility District, Wenatchee Washington.

Steig, T. W., and R. Adeniyi. 1995. Hydroacoustic evaluation of the behavior of juvenile salmon and steelhead approaching the powerhouse in the forebay of Rocky Reach Dam during 1995. Report by Hydroacoustic Technology, Inc. to Chelan County Public Utility District Number 1, Seattle.

Steig, T. W., and R. Adeniyi. 1996. Hydroacoustic evaluation of the fish passage through turbine units 1-11, spill gates 3-5, and the PSC at Rocky Reach Dam in the spring and summer of 1996. Report by Hydroacoustic Technology, Inc. to Chelan County Public Utility District Number 1, Seattle.

Steig, T. W., and S. V. Johnston. 1996. Monitoring fish movement patterns in a reservoir using horizontally scanning split-beam techniques. ICES Journal of Marine Science 53:435–441.

Steig, T. W., and T. K. Iverson. 1998. Acoustic monitoring of salmonid density, target strength, and trajectories at two dams on the Columbia River, using a split-beam scanning system. Fisheries Research 35:43–53.

Truscott, K. B., and S. G. Hays. 1988. Rocky Reach prototype fish guidance system 1987 developmental testing. Chelan County Public Utility District, Wenatchee, Washington.

Truscott, K. B., and S. G. Hays. 1989. Rocky Reach prototype fish guidance system 1988 developmental testing. Chelan County Public Utility District, Wenatchee, Washington.

Noah S. Adams, Dennis W. Rondorf, Scott D. Evans, Joe E. Kelly, Russell W. Perry, John M. Plumb, and Daniel R. Kenney

Migrational Characteristics of Radio-Tagged Juvenile Salmonids during Operation of a Surface Collection and Bypass System

7

A surface bypass collector (SBC) system has been identified as a potential means to improve downstream passage of juvenile salmonids through hydroelectric dams, thereby assisting the recovery of endangered salmon populations in the Columbia River basin. During the spring of 1996, 376 juvenile chinook salmon, 220 hatchery steelhead, and 168 wild steelhead were implanted with radio transmitters and released into Lower Granite Reservoir (Snake River, Washington) to evaluate a prototype SBC at Lower Granite Dam. About 20% of the fish came within 6 m of the SBC. Of these fish, about 45% (which is 9% of all fish detected in the forebay) passed the dam via the SBC. Average flow during the study was relatively high (3,514 m^3/s) compared to previous years (1994, 2,210.4 m^3/s; 1995, 2,805.5 m^3/s), and about 3% of the flow passed through the SBC. High flows likely accounted for about 80% of the fish passing the dam through the spillway or turbine intakes. Our results validated the concept of surface bypass; more information is needed, however, to determine how modifications to the SBC or changes in the operation of the powerhouse and spillway might increase the number of juvenile fish passing through the SBC. Because our results revealed species-specific differences in approach to and passage through the SBC, it is important to consider species composition, out-migration life stage, and origin when developing surface bypass systems.

Introduction

Years of research have been allocated to ensure the long-term survival of salmon and steelhead stocks in the Columbia River basin. Much of this effort has focused on the effects of dams and reservoirs on juvenile salmonids as they migrate from their natal waters to the ocean. Raymond (1968, 1979) and Park (1969) showed that dams increased migration times and suggested that this may be detrimental to juvenile salmon survival. Since the completion of Lower Granite Dam on the Snake River, Washington, delays of over 20 d have been observed (Buettner and Nelson 1989, 1990; Buettner 1991; Buettner and Brimmer 1993; Rondorf and Banach 1996), and some fingerlings, especially steelhead, have residualized in Lower Granite Reservoir during low flow years (Bennett 1992).

Research conducted in 1994 (Rondorf and Banach 1996) and 1995 (Adams et al. 1998) showed that radio-tagged juvenile chinook salmon and steelhead traveled through the Snake River and Lower Granite Reservoir quickly compared to long residence times in the forebay of Lower Granite Dam. This suggested that juvenile salmonids have more difficulty passing through the dam than traveling through the reservoir.

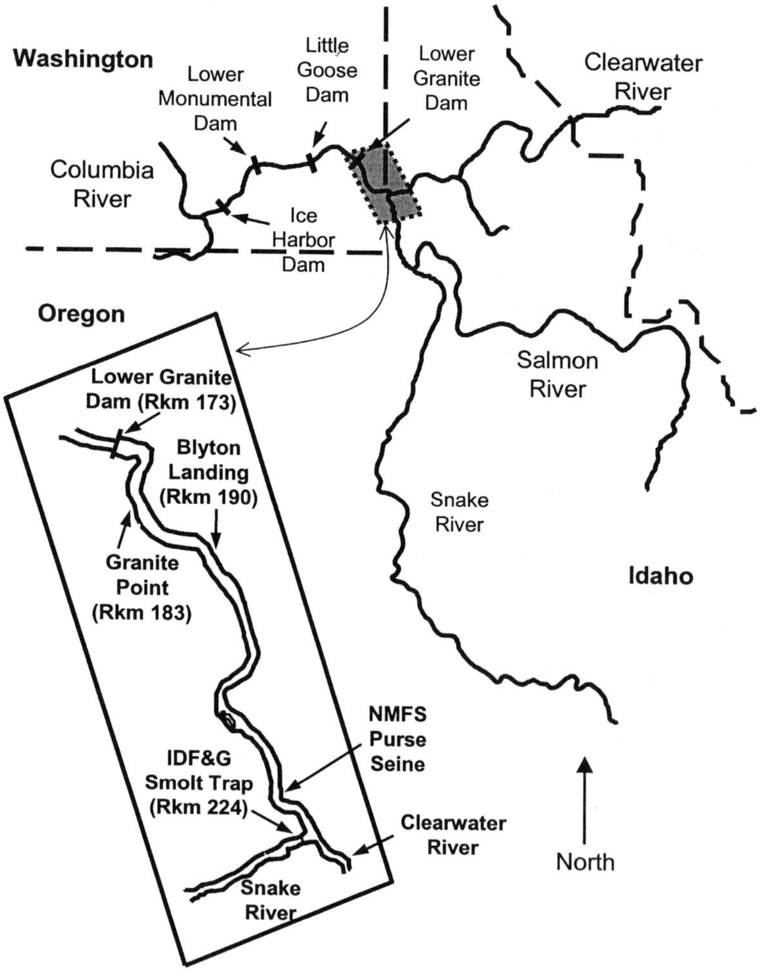

Figure 1. Map of Snake River drainage showing Lower Granite Reservoir study site for release of radio-tagged juvenile chinook salmon and steelhead during spring, 1996.`

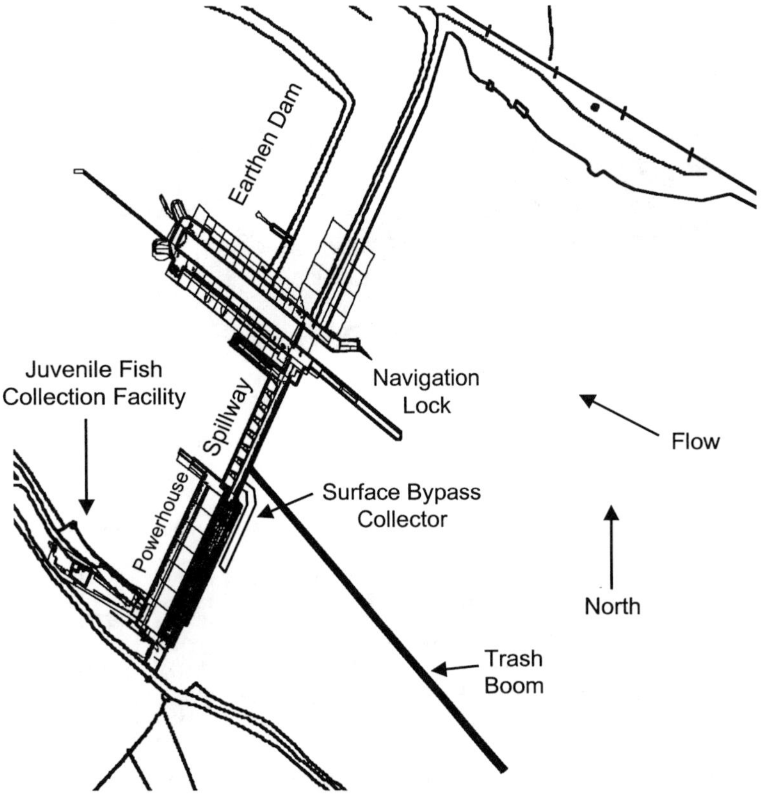

Figure 2. Schematic of Lower Granite Dam on the Snake River, Idaho, showing the main structural components of the dam and relative location of the surface bypass collector.

Reservoir drawdown, flow augmentation, spill, improved fish collection, and transportation systems have been identified as some of the management actions that may improve juvenile salmonid passage and survival, and thereby assist in the recovery of endangered salmon stocks in the Snake River. One option now being evaluated is surface bypass collection. In 1996, a prototype surface bypass collector was installed at Lower Granite Dam.

We used biotelemetry techniques to determine how effective the surface collector was in passing fish through Lower Granite Dam. Biotelemetry is an effective means of obtaining fish movement information (Winter 1983) and recent miniaturization of radio transmitters has made it possible to monitor juvenile salmonids as small as 120 mm fork length (Adams et al. 1998a, 1998b).

Methods

Study area and design

At full pool, Lower Granite Reservoir extends from Lower Granite Dam at river kilometer (RKM) 172 upriver to RKM 235 on the Snake River near Asotin, Washington (Figure 1). The dam is comprised of a six-unit powerhouse extending from the south shore, an eight-bay tainter gate controlled spillway near mid-river, and an earthen section extending from the north shore (Figure 2). Our study area extended from the dam upriver to Blyton Landing (RKM 190), and the forebay was defined as the area from the dam to about 3.5 km upriver.

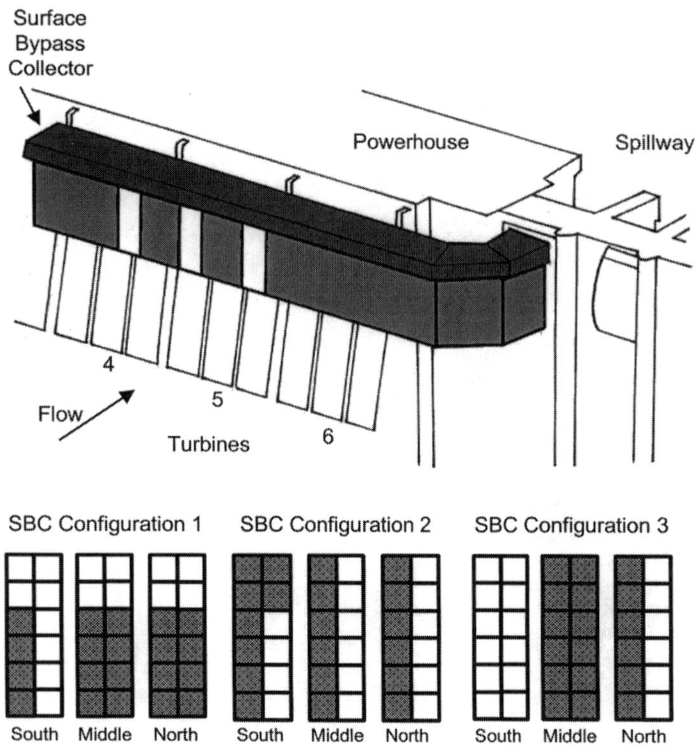

Figure 3. Three-dimensional representation of the surface bypass collector (SBC) retrofitted onto Lower Granite Dam in 1996 and the three opening configurations tested. Total discharge through the SBC when the doors were in configuration 1 and 2 was 110.5 m^3/s and 59.5 m^3/s for configuration 3. Shaded areas represent closed doors.

The SBC was retrofitted onto the face of the dam upstream of Turbines 4–6 at the north half of the powerhouse (Figure 3). The structure was a steel box 18 m high, 6 m wide, and 100 m long. Large flotation chambers at the top of the SBC allowed it to move vertically with changes in forebay elevation. The bottom of the SBC was 17 m from the water surface. There were three entrances to the SBC and each entrance had six sets of sliding doors. With the doors fully open, each of the three entrances was about 5 m wide by 15 m high. Different entrance configurations were tested by opening and closing individual doors (Figure 3).

The SBC configurations were designed to test the passage efficiencies of horizontal and vertical entrance orientations, maximum and reduced entrance area, as well as high and low water velocity entrances. Area and velocity effects were confounded because, for a given SBC discharge, one factor could not be changed without affecting the other. Configurations were sampled according to a randomized block design. There were ten 3-d blocks (one SBC configuration tested per day).

Test fish and tagging

We tagged and released 376 juvenile hatchery chinook salmon, 220 juvenile hatchery steelhead, and 168 juvenile wild steelhead. Most of the test fish were obtained from Idaho Department of Fish and Game's smolt monitoring trap (RKM 224) and the National Marine Fisheries Service purse seine operation (RKM 220). Additional fish were seined along the banks of the Snake River near Blyton Landing and at a site 1 km upstream from the confluence of the Snake and Clearwater Rivers. To allow fish to adjust to migration in the reservoir before reaching the dam, fish were tagged and released at Blyton Landing, about 20 km upstream of the dam.

Fish were transferred from collection sites to net pens at Blyton Landing and held in the river for at least 24 h before tagging. All fish were surgically implanted with coded radio transmitters (Lotek Engineering Inc., Newmarket, Ontario, Canada) using procedures described by Adams et al. (1998a, 1998b). Releases occurred in the morning (about 0800 hours) and afternoon (about 1600 hours) to stagger fish arrival time at the dam. We used newly developed radio tags that allowed multiple tags to broadcast on the same frequency. These coded tags allowed us to monitor 100 tags broadcasting on the same frequency without losing the ability to identify individuals. By using only eight frequencies to monitor the movements of nearly 800 fish, the receiver's scan cycle was greatly reduced and the probability of not detecting a tagged fish was low. To further enhance our detection capabilities, we combined receivers with digital spectrum processor (DSP) units. Because a DSP listens to all frequencies and codes simultaneously, there is no scan cycle and all fish within range of a DSP's antenna can be detected nearly instantaneously.

Data Collection

Fifty-five fixed aerial antennas linked to 12 data logging receivers (Lotek Engineering Inc.), and 71 underwater antennas (designed by Grant Systems Engineering, Newmarket, Ontario, Canada) linked to 12 data logging receivers were used to monitor fish movements within the forebay of Lower Granite Dam 24 h/d, 7 d/week. Aerial arrays were installed at eight locations: Granite Point (located 10 km upstream of Lower Granite Dam), forebay barge sites (located 0.8 km upstream of Lower Granite Dam), earthen dam, logboom, spillway, powerhouse, juvenile fish collection facility (JFCF), and exit sites (Figure 2; located 0.8 km downstream of Lower Granite Dam). Each aerial array consisted of multiple antennas linked to a Lotek SRX 400-W16 receiver. The receiver was programmed to scan eight frequencies for 3 s each, resulting in a 24 s scan cycle. Underwater arrays were installed in four locations: spillway, powerhouse, surface bypass collector, and fish gallery (Figures 2 and 3). The fish gallery is a concrete raceway located inside the dam that runs the length of the powerhouse. Fish that enter the turbine intakes and encounter the extended length bar screens (ELBS) are diverted into the fish gallery. Once inside the fish gallery, they travel through underground pipes to the JFCF. Each underwater array consisted of multiple antennas linked to a Lotek SRX 400-W16 receiver/DSP unit.

Field studies before releasing test fish revealed that the detection range of aerial antennas varied from 130 to 300 m depending on the depth at which radio-tagged fish traveled. Fish traveling in the top 3 m of the water column could be detected from 200 to 300 m, whereas fish traveling between 3 and 10 m deep could be detected from 130 to 240 m. The physical structure of the underwater antennas together with the characteristics of underwater signal propagation and reception limited the range of detection to about 10 m. Because the strength of the signal increases as distance between the tagged fish and the antenna decreases, signal strength was used to determine the depth at which fish approached the SBC. Similarly, the area of the dam where fish were first detected was used to determine its approach path.

Data collection began on 23 April 1996 and continued until 15 June 1996 (19 d after the last release) when it was determined that there were no test fish within the study area. Residence times in the forebay (first detection to last detection at the dam) were determined by summing the number of hours a test fish was in the forebay. Residence times were only calculated for fish that were detected in the forebay and then below the dam (i.e., confirmed passing the dam). The records for these fish were then scrutinized to determine routes of passage through the dam. There were four passage routes through Lower Granite Dam: the SBC, through the spillway, into the turbine intakes and through the turbine unit, or into the turbine intakes and diverted by the ELBS into the JFCF.

Results

Mortality and detection efficiency

Thirty (7% of total fish tagged) fish died after tagging and before release into the reservoir. Eight of the 30 mortalities were caused by tangling of the radio antennas while fish were held 24 h before release. The remaining 22 fish died of unknown causes.

Ninety-one percent of the radio-tagged chinook salmon, 98% of the hatchery steelhead, and 100% of the wild steelhead were detected two or more times by fixed-site receiving stations. About 70% of all the fish were detected more than 100 times. The majority of first detections occurred in the earthen dam area. Fifty-four percent of hatchery chinook salmon, 59% of hatchery steelhead, and 61% of wild steelhead were first detected by the earthen dam antenna array.

Residence time

The median residence time in the forebay for hatchery chinook salmon (1.9 h) was greater than the median residence time for hatchery steelhead (0.7 h) and wild steelhead (1.0 h). There was no clear relation between residence time and total discharge, spill, or percent of total discharge spilled. The lack of any clear relation between discharge and residence time was likely due to high flows in the Snake River and continuous spill at the dam during the entire study period (Figure 4). However, there was a relation between residence time and route of passage for fish that came within 10 m of the SBC. Fish that came within 10 m of the SBC but did not enter had longer median residence times (chinook salmon 3.7 h; hatchery steelhead 2.3 h; wild steelhead, 2.9 h) than fish that entered and passed through the SBC (chinook salmon 1.3 h; hatchery steelhead 0.8 h; wild steelhead, 1.9 h).

Fish movements in forebay of Lower Granite Dam

The spatial distribution of fish detected at each of the eight aerial arrays was similar for hatchery chinook salmon, hatchery steelhead, and wild steelhead. Among the aerial arrays in the immediate forebay of the dam, the logboom and powerhouse arrays detected 32–44% of all the radio-

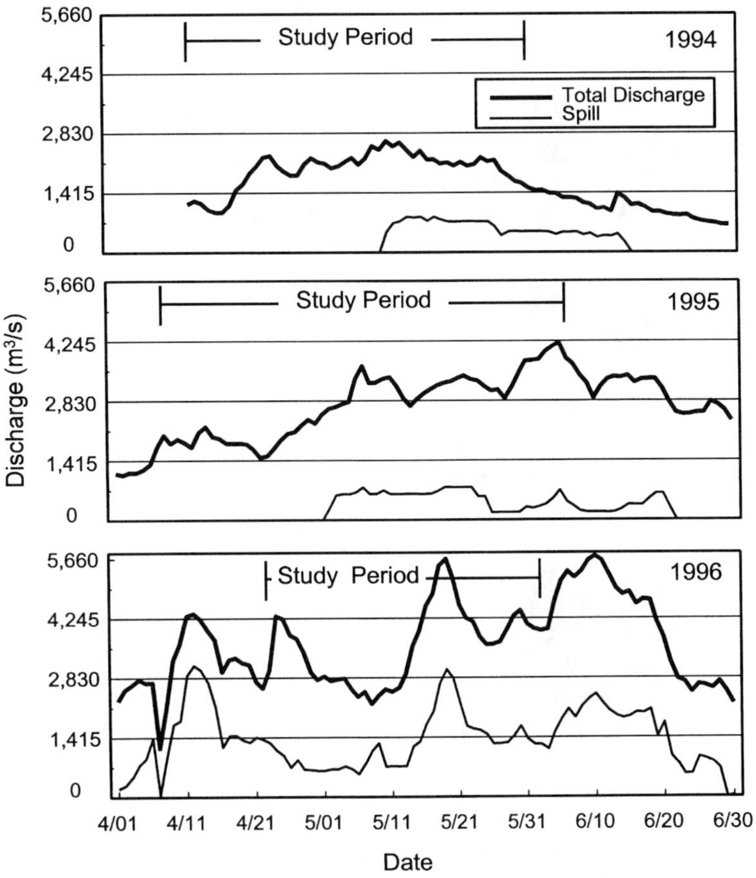

Figure 4. Average daily total discharge and spill at Lower Granite Dam in 1994, 1995, and 1996 relative to when radio-tagged juvenile salmonids were released into the study area.

tagged fish and the spillway and earthen dam arrays detected 56–81% of all radio-tagged fish. Time spent in a particular area of the forebay differed between chinook salmon and steelhead. Based on the percent of the total detections recorded by each of the eight aerial arrays, we estimated that radio-tagged chinook salmon spent most of their time in the powerhouse and earthen dam areas, and hatchery and wild steelhead spent most of their time in the spillway and earthen dam areas. Within the spillway and powerhouse arrays, number of detections at individual antennas indicated that radio-tagged fish spent most of their time in the area where the powerhouse and spillway intersect. This was the same general area where the SBC was located.

Passage through Lower Granite Dam

Passage routes through the dam were determined for 80% of radio-tagged chinook salmon, 95% of hatchery steelhead, and 89% of wild steelhead. Depending on species, 68–80% of radio-tagged fish passed through Lower Granite Dam via the spillway and juvenile fish collection system (Figure 5). Passage routes varied according to the way fish approached the dam. Seventy-four percent of the chinook salmon, 88% of the hatchery steelhead, and 84% of the wild steelhead that were first detected in the earthen dam area passed the dam through the spillway. Fish that were first de-

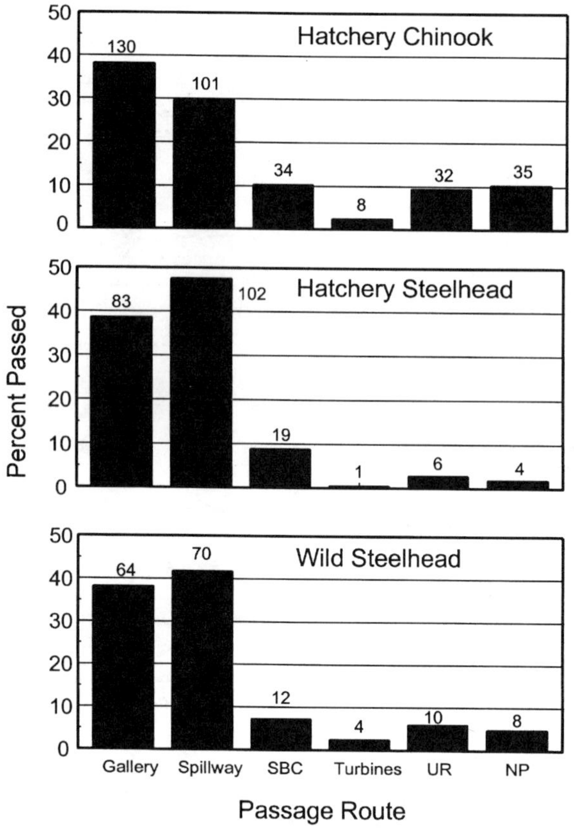

Figure 5. Routes of passage through Lower Granite Dam for radio-tagged chinook salmon, hatchery steelhead, and wild steelhead. (UR, fish that passed the dam through and unknown route; NP, fish that were detected in the forebay but where not detected passing through the dam).

tected in the spillway area passed the dam in relatively equal proportions through the spillway (43% hatchery chinook salmon, 36% hatchery steelhead, 36% wild steelhead) and JFCF (33% hatchery chinook salmon, 42% hatchery steelhead, 44% wild steelhead). Eighty-four percent of the hatchery chinook salmon, 65% of the hatchery steelhead, and 76% of the wild steelhead that were first detected in the area in front of the south side of the powerhouse passed through the dam by entering the turbine intakes and were then guided into the JFCF. Similarly, 50% of the hatchery chinook salmon, 58% of the hatchery steelhead, and all of the wild steelhead that were first detected near the SBC passed the dam through the JFCF. Passage into the JFCF occurred mainly through turbine intake 5, directly under the SBC. The time between detection at the turbine intake entrance and detection in the juvenile fish collection gallery for chinook salmon (mean = 12 min) and steelhead (mean = 17 min) suggested that fish did not delay in the gatewells before moving into the fish gallery.

Depending on species, 7–11% of the radio-tagged fish passed the dam via the SBC. About 20% of all radio-tagged fish came within 10 m of the openings to the SBC. Of these fish, 37–51% passed into and through the SBC. Most of the radio-tagged chinook salmon (46%) passed into the SBC

Table 1. Number of radio-tagged juvenile salmonids passing into the surface bypass collector (SBC) during three different opening configurations in 1996.

	SBC opening configuration		
Species	1	2	3
Chinook salmon	16	12	7
Hatchery steelhead	5	9	5
Wild steelhead	2	6	4

during configuration #1 whereas most of the hatchery (47%) and wild (51%) steelhead passed into the SBC during configuration #2 (Table 1). Of the fish that came within 10 m of the openings to the SBC but did not enter, 83–88% passed under the collector and into the JFCF via turbine intakes 4, 5, and 6.

The depth at which fish approached the SBC affected whether fish eventually went in, under, or around the SBC. For chinook salmon and hatchery steelhead, the greatest number of fish that approached the SBC at a depth less than 10 m passed into and through the SBC (Table 2). For wild steelhead that approached at a depth less than 10 m, the greatest number of fish passed under the SBC and into turbine intakes. In contrast, most fish (regardless of species) that approached the SBC at depths more than 10 m passed under it (Table 2).

The depth at which fish entered the SBC varied between species. About half (47%) of the hatchery chinook salmon entered the SBC toward the top of the openings in the collector (0–5 m). Hatchery steelhead entered the collector in about the same proportion regardless of depth, and 66% of the wild steelhead entered through the middle of the openings to the collector (5–10 m).

Regardless of species, fish were inside the SBC for about 2 min before passing through, and no fish stayed inside the collector for more than 10 min. There was no obvious diel pattern for passage into the SBC. However, there was a tendency for fish to enter the JFCF during nighttime hours. There was no apparent relationship between the occurrence of spill and the number of fish that passed into the SBC or the JFCF.

Table 2. Number of radio-tagged juvenile salmon that passed in, under, or around the surface bypass collector (SBC) relative to the depth at which they approached the SBC.

	Approach depth < 10 m			Approach depth > 10 m		
Species	Into SBC	Under SBC	Around SBC	Into SBC	Under SBC	Around SBC
Chinook salmon	25	10	6	10	14	1
Hatchery steelhead	15	12	4	5	11	0
Wild steelhead	11	15	2	1	3	0

Discussion

Even though the flow passing through the SBC represented less than 3% of the total flow through Lower Granite Dam, 7–11% of all radio-tagged fish detected in the forebay passed through the dam via the SBC. Although these results validate the concept of surface collection at Lower Granite Dam, the effectiveness of the surface collector in passing juvenile salmonids was low compared to surface collection systems at other locations (Johnson et al. 1992; T. W. Steig and B. H. Ransom, Hydroacoustic Technology, Inc., Seattle, unpublished data). In a review of surface spill and deep spill, Ransom and Steig (unpublished data) estimated sluiceways typically produced a 13:1 ratio of the percentage of total fish to percentage of total river flow passed. In contrast, the deep flow of conventional spillways passed fish at about a 1:1 ratio of fish to flow.

Although only a small percentage of the total number of radio-tagged fish passed through the SBC, this number represented 37–51% of all the radio-tagged fish that came within 10 m of the openings to the SBC. These data suggested that once fish were close enough to discover the flow going into the SBC, about half the fish entered the SBC. Sample size was too small to determine if SBC configuration had a significant effect on the number of fish moving into the collector. Our data are supported by fixed hydroacoustic data which indicated that 39–45% of the fish that came within 3 m of the SBC passed into the SBC (Johnson et al. 1996). Results of studies conducted at other dams within the Columbia River Basin (reviewed by Giorgi and Stevenson 1995) also offer evidence that flow nets near the surface may be effective at passing juvenile salmonids.

We do not know if fish detected flow moving into the SBC and then actively chose to enter the SBC. It is also possible that fish randomly moved into the SBC depending on their position in the water column. Irrespective of whether fish moved into the SBC actively or passively, data on fish distribution in the forebay of Lower Granite Dam suggested that more fish should have moved through the SBC than were reported. For example, mobile hydroacoustic surveys conducted 15 m upstream of the SBC revealed that 80% of the fish were in the upper 18 m and 50% of the fish were in the upper 10 m of the water column (Kofoot et al. 1996). Furthermore, hydroacoustic data showed that fish density was consistently higher in the area in front of the SBC compared to the area in front of the spillway, earthen dam, and south side of the powerhouse. Additionally, radiotelemetry data collected since 1994 showed that fish congregated at the intersection of the spillway and powerhouse where the SBC is currently located.

The hydraulic environment in front of and under the SBC may account for the relatively low number of fish passing into the SBC. A physical model of Lower Granite Dam (U.S. Army Corps of Engineers, Waterways Experiment Station, Vicksburg, Mississippi) together with data obtained with an acoustic Doppler current profiler (ADCP; Kofoot et al. 1996) revealed strong downward movement of water in front of the SBC. In addition, starting about 30 m upstream of the SBC, most of the water flowed under the SBC and into turbine intakes 4, 5, and 6. It is likely that most of the fish in the vicinity of the SBC became entrained in the flow of water going into the turbines and were never exposed to the entrances to the SBC (Johnson et al. 1998). Thus, it is reasonable to assume that if powerhouse operations were altered so that less water flowed under the SBC, more fish may have passed into the SBC.

Passage into the SBC may have been influenced by the path fish took when approaching the dam. Because most of the fish approached the dam along the north shore and then moved south across the forebay, many of them may have passed through the spillway before encountering the SBC. The relatively low number of fish detected in the area in front of the powerhouse indicated that lateral movement across the face of the dam by radio-tagged fish was limited during 1996 compared to results from 1994 and 1995 (Rondorf and Banach 1996). Previously, fish approached the dam down the middle of the river and along the south shore. Had this been the case during 1996, it is likely that more fish would have encountered and potentially passed into the SBC. Differences in approach paths between years is best attributed to the high flows in the Snake River, and resulting spill at the dam, during 1996 (Figure 4).

Increased spill due to high flows likely accounted for decreased forebay residence times. The median forebay residence times for all species in 1996 were less than the median forebay residence times in 1994 (chinook salmon, 5.0 h; hatchery steelhead, 26.7 h; wild steelhead, no data) and 1995 (chinook salmon, 6.8 h; hatchery steelhead, 14.2 h; wild steelhead, 1.1 h).

In summary, our results showed that surface collection may be an effective way to pass juvenile salmonids through Lower Granite Dam. We feel that modifications to the SBC and changes in the operation of the powerhouse could increase the number of fish passing through the SBC. However, the relatively low numbers of fish entering the SBC despite relatively high abundance upstream indicated a need for a better understanding of fish behavior near surface bypass entrances. Because fish tend to pass Lower Granite Dam quickly during periods of high discharge, it may not be necessary to utilize this passage alternative under these conditions. However, low flows increase fish residence time in the forebay, and our data indicate that surface collection has the potential to substantially reduce fish delay in the forebay of Lower Granite Dam.

Acknowledgments

Funding for this study was provided by the U.S. Army Corps of Engineers, Walla Walla District, Walla Walla, Washington (contract E-86930151). The use of trade names does not imply endorsement of commercial products.

References

Adams, N. S., D. W. Rondorf, M. A. Tuell, and M. J. Banach. 1998. Migrational characteristics of juvenile spring chinook salmon and steelhead in Lower Granite Reservoir and tributaries, Snake River. Report to the U.S. Army Corps of Engineers, Contract E-86930151, Walla Walla, Washington.

Adams, N. S., D. W. Rondorf, S. D. Evans, and J. E. Kelly. 1998a. Effects of surgically and gastrically implanted radio transmitters on growth and feeding behavior of juvenile chinook salmon. Transactions of the American Fisheries Society 127:128–136.

Adams, N. S., D. W. Rondorf, S. D. Evans, J. E. Kelly, and R. W. Perry. 1998b. Effects of surgically and gastrically implanted radio transmitters on swimming performance and predator avoidance of juvenile chinook salmon (*Oncorhynchus tshawytscha*). Canadian Journal of Fisheries and Aquatic Sciences 55:781–787.

Bennett, D. H. 1992. Residualism of salmonid fishes in Lower Granite Reservoir, Idaho-Washington. Pages 81–86 *in* Passage and survival of juvenile chinook salmon migrating from the Snake River basin. University of Idaho, Moscow.

Buettner, E. W. 1991. Smolt monitoring at the head of Lower Granite Reservoir and Lower Granite Dam. Annual report for 1991 operations to Bonneville Power Administration, Contract DE-BI79-83bp11631, Portland, Oregon.

Buettner, E. W., and V. L. Nelson. 1989. Smolt condition and timing of arrival at Lower Granite Reservoir. Annual report for 1988 operations to Bonneville Power Administration, Contract DE-AI79-83BP11631, Portland, Oregon.

Buettner, E. W., and V. L. Nelson. 1990. Smolt monitoring at the head of Lower Granite Reservoir and Lower Granite Dam. Annual report for 1989 operations to Bonneville Power Administration, Contract DE-AI79-83BP11631, Portland, Oregon.

Buettner, E. W., and A. F. Brimmer. 1993. Smolt monitoring at the head of Lower Granite Reservoir and Lower Granite Dam. Annual report for 1992 operations to Bonneville Power Administration, Contract DE-BI79-83BP11631, Portland, Oregon.

Kofoot, E. E., M. E. Hanks, and J. B. Oleyar. 1996. Distribution of juvenile salmonids in the forebay of Lower Granite Dam detected during mobile hydroacoustic surveys. Pages 135–183 *in* N. S. Adams, D. W. Rondorf, E. E. Kofoot, and M. A. Tuell, editors. Migrational characteristics of juvenile spring chinook salmon and steelhead in the forebay of Lower Granite Dam relative to the 1996 surface bypass collector tests. Report to the U.S. Army Corps of Engineers, Contract E-86930151, Walla Walla, Washington.

Giorgi, A. E., and J. R. Stevenson. 1995. A review of biological investigations describing smolt passage behavior at Portland District Corps of Engineers projects: implications to surface collection systems. Report of Don Chapman Consultants to U.S. Army Corps of Engineers, Portland, Oregon.

Johnson, G. E., and eight coauthors. 1998. Fixed-location hydroacoustic evaluation of the prototype surface bypass and collector, spill efficiency, and fish guidance efficiency at Lower Granite Dam in spring and summer 1997. Report to the U.S. Army Corps of Engineers, Contract DACW68-96-D-0002, Walla Walla, Washington.

Johnson, G. E., C. M. Sullivan, and M. W. Erho. 1992. Hydroacoustic studies for developing a smolt bypass system at Wells Dam. Fisheries Research 14:221–237.

Park, D. L. 1969. Seasonal changes in downstream migration of age-group 0 chinook salmon in the upper Columbia River. Transactions of the American Fisheries Society 98:315–317.

Raymond, H. L. 1968. Migration rates of yearling chinook salmon in relation to flows and impoundments in the Columbia and Snake rivers. Transactions of the American Fisheries Society 97:356–359.

Raymond, H. L. 1979. Effects of dams and impoundments on migrations of juvenile chinook salmon and steelhead from the Snake River, 1966 to 1975. Transactions of the American Fisheries Society 108:505–529.

Rondorf, D. W., and M. J. Banach. 1996. Migrational characteristics of juvenile spring chinook salmon and steelhead in Lower Granite Reservoir and tributaries, Snake River. Report to the U.S. Army Corps of Engineers, Contract E 86930151, Walla Walla, Washington.

Wilson, J. W., A. E. Giorgi, and L. C. Stuehrenberg. 1991. A method for estimating spill effectiveness for passing juvenile salmon and its application at Lower Granite Dam on the Snake River. Canadian Journal of Fisheries and Aquatic Sciences 48:1872–1876.

Winter, J. D. 1983. Underwater biotelemetry. Pages 371–395 *in* L. A. Nielsen and D. L. Johnson, editors. Fisheries techniques. American Fisheries Society, Bethesda, Maryland.

Dilip Mathur, Paul G. Heisey, John R. Skalski, and Daniel R. Kenney

Survival of Chinook Salmon Smolts through the Surface Bypass Collector at Lower Granite Dam, Snake River

8

Surface bypass collection systems (SBC) are increasingly relied upon in the Pacific Northwest to take advantage of the surface-oriented behavioral pattern of young salmonids in diverting them away from the hydro turbines. The newly installed prototype surface collector at Lower Granite Dam, Snake River, has the potential to pass the highest number of drainage smolts listed under the Endangered Species Act. Survival and condition of hatchery-reared chinook salmon *Oncorhynchus tshawytscha* (132–183 mm, average about 149 mm total length) in passage through the SBC were estimated using the balloon tag–recapture method. Because the fish diverted by the SBC go over a spillbay, their survival was also estimated to isolate the effects of spillbay passage from that through the SBC. Fish exposed to SBC suffer mortality from two sources: SBC itself and spillbay. Relative to tailrace controls, the 48 h survival probability for the SBC fish was estimated at 0.958 (90% confidence interval [CI] = 0.928–0.988) and it was 0.983 (90% CI = 0.941–1.0) relative to spillbay releases. Relative to the tailrace controls, the estimated survival probability for spillbay fish was 0.975 (90% CI = 0.951–0.988). The overall estimated passage survival for the SBC is within range of that reported for other exit routes including some hydro turbines.

Introduction

Juvenile salmonids migrating downstream in the Columbia River Basin encounter any or all of the following exit routes at hydroelectric dams: turbines, spillways, sluiceways, and bypasses. There are two interrelated issues associated with passage through any of these routes. One is the proportion of fish utilizing any of these routes during emigration. The second is the fish condition or survival subsequent to passage. Spill has been used to enhance overall survival of juvenile emigrants because of reported high survival (≥98%) and greater effectiveness at hydroelectric dams in the Columbia River Basin (Schoeneman et al. 1961; Heinle and Olson 1981; Ledgerwood et al. 1990). However, spill is expensive in terms of lost power generation and can result in potentially lethal levels of total dissolved gas in the river. Thus, there has been a need to evaluate alternative means to divert juvenile salmonids away from turbines. Also, because surface-oriented salmonid emigrants generally occupy the top 10 m (Raemhild et al. 1985) of the water column, they must sound up to 20 m to exit the bottom-opening tainter gates at spillways. To take advantage of the surface-oriented behavioral characteristic, the emigrating smolts may be intercepted in a surface bypass collection system (SBC) to safely bypass the turbine passage route. However, survival and fish condition after passage through a SBC are unknown.

Johnson (1996) described the successful application of a surface collection system at Wells Dam on the Columbia River to bypass salmonid emigrants. The U.S. Army Corps of Engineers have been evaluating the effectiveness of the SBC prototype, retrofitted to Lower Granite Dam (LGD) on the Snake River (river km 161). The SBC was designed and built to safely intercept emigrating salmonid smolts before turbine entrainment or diversion by fish guidance intake screens. The SBC at LGD has the potential to collect a high proportion of salmon smolts listed under the Endangered Species Act since LGD is the most upstream dam on the Lower Snake River.

Although it was hypothesized that the injury or mortality due to passage through the SBC would be low, no empirical assessment had been made. Consequently, the objective of the present study was to evaluate the direct effects of passage through the SBC prototype structure on the physical condition and survival of hatchery-reared yearling chinook salmon *Oncorhynchus tshawytscha*.

Site Description

Lower Granite Dam is the fourth upstream dam on the Snake River, near Almota, Washington. It was placed in service in 1975. The powerhouse is about 197 m long, has an operating head of about 29 m, and houses six vertical shaft Kaplan propeller-type turbines. Each turbine has three bays with an extended-length screen in each bay.

The SBC structure is in front of turbine Units 4, 5, and 6 and is connected to Spillbay 1 (Figure 1). Three entrances to the structure exist, each with six pairs of gates, which can be pneumatically operated independently. Up to about 119 m^3/s of water passes through the SBC to the spillbay. The tainter gate on Spillbay 1 controls the total volume of flow through the SBC, but the 12 gates at each entrance structure control the

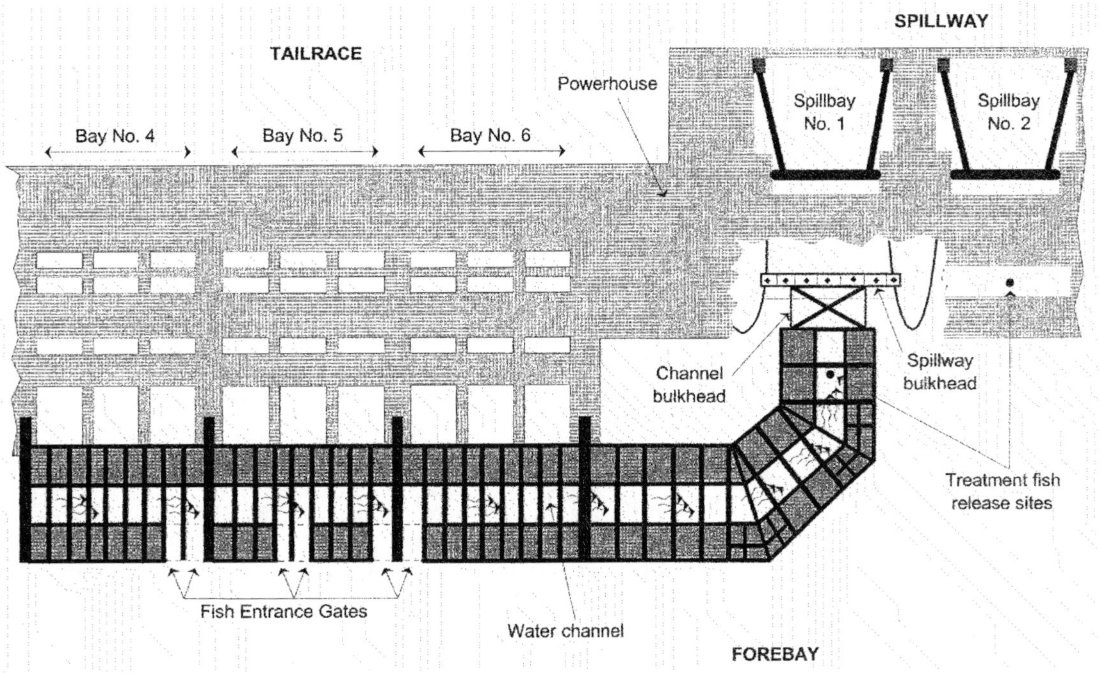

Figure 1. Schematic top view of the surface bypass collector (SBC) at Lower Granite Dam and fish release locations.

volume of water passing the three entrances. Entrance gate velocity can be varied from less than 0.3 m^3/s to more than 2 m^3/s, depending on the combination of entrance gate and tainter gate positions. The entrance and tainter gate openings can also vary the mean velocity in the main channel of the structure from about 0.7 to 1.2 m^3/s. Close to the tainter gate, water velocity increases up to about 13 m^3/s. Fish diverted by SBC pass under the Spillbay 1 tainter gate and are entrained in its flow, similar to fish transported through other spillbays. Nine 0.3 m orifices (to allow fish an escape route from behind the structure) also exist on the downstream side of the SBC, but make a negligible contribution to total flow volume.

Methods

Sample size

Fish were allocated to three release locations (Figure 1): SBC (N = 120) to delineate the effects of SBC; an adjacent Spillbay 2 (N = 120) to provide information on the magnitude of mortality associated with spillbay passage; and the tailrace (N = 100) to account for the effects of handling, tagging, induction, and recapture. We calculated that a precision (ϵ) of at least ± 0.05 was possible 90% of the time if the recapture and control survival rates equal or exceed 0.95 with the experimental protocol employed.

Experimental fish were transported about 400 km from the Little White Salmon National Hatchery, Washington. At the study site, fish were held in two circular tanks, each with 900 L capacity. Lots of 100–150 fish were transferred to a 900 L tank on the upper deck of the powerhouse (treatment release) and another tank on the lower gallery (control release).

Fish were held for about 24 h before testing; holding tanks were continuously supplied with ambient river water. Treatment and control fish were drawn from the same group of fish assuring similar size and condition. The average total lengths for the three test groups ranged from 148 to 150 mm.

Tagging and release

Balloon tagging and release techniques were identical to those used elsewhere for juvenile salmonids (Mathur et al. 1996). To track the 48 h survival of individual fish, each fish was also equipped with a uniquely numbered visual implant (VI) tag (Northwest Marine Technology, Inc., Shaw Island, Washington). Fish were allowed to recover from anesthesia in a covered 90 L container (generally 40–70 min), continually supplied with ambient river water, before release through the induction apparatus. Upon full recovery from anesthesia, fish were individually placed into the induction system, tags activated, and released within a continuous flow of water (Heisey et al. 1992).

Fish in the SBC were released at a high velocity (about 1.1–1.9 m/s) location near the center of the channel where they were fully committed to downstream passage. The terminus of the hose was secured with a weight (approximately 18 kg) and three ropes so that fish were discharged with the flow about 6 m below the surface and 6 m from the downstream end of the SBC. The tailrace control fish were released into the Spillbay 1 discharge (Figure 1). The flexible release hose was inserted into a 13 cm rigid PVC (polyvinyl chloride) pipe and extended about 4.2–6 m off the north end of the powerhouse gallery.

For the Spillbay 2 releases, the terminus of the induction hose passed through a 15 cm sweep elbow, suspended by a crane, to ensure that the hose remained at the desired location. The flexible hose was attached to the supporting crane cable along its length from the powerhouse deck level to the sweep elbow. Guy wires from the weighted sweep elbow to locations on the spillbay nose piers insured that the induction hose was oriented downstream.

We utilized a single tailrace control release for both the SBC and spillbay-exposed smolts. This experimental protocol has proven efficient, particularly when the fish availability is low (Mathur et al. 1997). We alternately released 15–20 fish through the SBC and spillbay followed by a release of a similar number of controls on each day. Fish were released through identical induction systems at the three locations, with approximately 114 m^3/s simultaneously discharged from Spillbays 1 and 2. Owing to higher than expected river flow, total spill ranged from 349 to 1,140 m^3/s during the fish releases. Most of the time, spill rate was near 228 m^3/s.

Fish recapture

Shortly after release, the tags inflated and buoyed the fish to the surface for rapid recapture (generally less than 10 min). Fish were retrieved and placed into an onboard holding facility. The tags were removed, and each fish was examined for descaling and injuries and transferred in 5 gallon pails to an onshore holding pool for 48 h observations.

Classification of recaptured fish

Recaptured fish and recovery of inflated dislodged tags were classified per Mathur et al. (1996) to estimate the immediate (1h) and postpassage (48 h) effects. Mortalities occurring after 1h postpassage were considered 48 h mortalities. Dead fish were necropsied to determine the probable cause of death. Additionally, all specimens alive at 48 h were re-anesthetized and closely examined for injury and descaling. Injuries were recorded at the initial recapture and also at 48 h.

Data analysis

Data were analyzed using maximum likelihood model (Mathur et al. 1996) to estimate the parameters and their associated standard errors. However, because the two releases (SBC and Spillbay 2) were made concurrently with a single shared control group, the likelihood model, accounting for dependencies within the study design, was modified to include the two simultaneous treatments (Mathur et al. 1997). A *Z*-statistic was used to detect differences ($P = 0.05$) in survival probabilities between SBC and Spillbay 2 fish.

Survival probabilities were calculated for the SBC and Spillbay 2 fish relative to tailrace controls. Survival probability for the SBC fish was also calculated relative to survival of Spillbay 2 fish to isolate the effects of spill passage. Parameter estimates along with their standard errors were based on the simplified model ($H_O{:}P_A{=}P_D$), that is, equality in the recapture probabilities of alive and dead fish (Mathur et al. 1996).

Results

Recapture rates

Nearly all individuals of treatment and control groups were recaptured (Table 1). The recapture rates were 99.2% for the SBC group and 100% for the Spillbay 2 and control groups. The proportion of SBC and spillbay fish that were physically recaptured dead was 0.025 and 0.017, respectively (Table 1). No control fish were dead upon recapture or later. Tag dislodgment occurred in only one SBC fish, and it was assumed dead in the analysis.

Recapture times for the three release groups were short and similar, averaging less than 8 min. The difference in average recapture times between groups was less than 0.4 min.

Survival probabilities

Survival probabilities were relatively high for both the SBC and spillbay fish (Table 2). Relative to tailrace controls, the estimated immediate (1 h) survival probabilities were 0.967 (90% CI = 0.94–0.994) for the SBC fish and 0.983 (90% confidence interval [CI] = 0.945–1.0) for the Spillbay 2 fish. Relative to Spillbay 2, the estimated survival probability for the SBC fish was 0.984 (i.e., 0.967 divided by 0.983), with a 90% CI of 0.945–1.0. The difference of 0.016 in survival probability between the SBC and Spillbay 2 fish may be attributed to the SBC passage alone. Because of the small sample size, however, this difference was not significant ($P\,|\,Z\,|\,>$

Table 1. Tag–recapture data on chinook salmon smolt survival and condition estimation in passage through the surface bypass collector (SBC) and Spillbay 2 at Lower Granite Dam, April 1996. Proportions are given in parenthesis.

	Trial 1	Trial 2	Trial 3	Trial 4	Total
		SBC			
Number released	15	35	40	30	120
Number recaptured alive	15	34	38	29	116 (0.967)
Number recaptured dead	0	0	2	1	3 (0.025)
Dislodged tags[a]	0	1	0	0	1 (0.008)
Number alive at 48 h	14	34	38	29	115 (0.958)
		Spillbay 2			
Number released	15	35	35	35	120
Number recaptured alive	14	35	35	34	118 (0.983)
Number recaptured dead	1	0	0	1	2 (0.017)
Number alive at 48 h	13	35	35	34	117 (0.975)
		Control			
Number released	5	30	30	35	100
Number recaptured alive	5	30	30	35	100 (1.0)
Number alive at 48 h	5	30	30	35	100 (1.0)

[a] Fish assumed dead.

0.828 = 0.407). However, owing to near complete recapture rates (>99%) and ideal control survival of 100%, the sample size was sufficient to provide a precision (ϵ) of at least ±4%, 90% of the time on the survival estimates (Table 2).

Few mortalities (one additional fish in each group) occurred after 1 h (Table 2). The 48 h survival probabilities for the SBC fish, relative to tailrace controls, were estimated at 0.958 (90% CI = 0.928–0.988) and at 0.975 (90% CI = 0.951–0.998) for Spillbay 2 fish. Relative to Spillbay 2, the estimated survival for the SBC fish was 0.983 (90% CI = 0.941–1.0). The difference between the two groups at 48 h was 0.017, or essentially the same at 1 h, and was nonsignificant ($P \mid Z \mid > 0.720 = 0.41$).

Injury type

Two of 119 (1.7%) SBC fish were visibly injured while 1 of 120 (0.8%) Spillbay 2 fish recovered was injured (Table 3). One of 100 (1.0%) control fish was visibly injured. Thus, relative to controls and spillbay, the injury rate attributable to the SBC passage was negligible. However, the injury type incurred by the three groups appeared different (Table 3). Of the two SBC visibly injured fish, one had a hemorrhaged eye and a flap of skin peeled off the head, and the other had a hemorrhaged liver. The one Spillbay 2 injured fish showed kidney damage, internal bruising, and also was bent in half. All these fish with visible injuries died. The only control fish with a visible injury (bulging eye) was alive at 48 h.

Table 2. Estimated passage survival probabilities of chinook salmon smolts at the surface bypass collector (SBC) at Lower Granite Dam, April 1996. Ninety percent profile confidence intervals are shown in parentheses. Spillbay 2 fish survival probability was calculated relative to tailrace control releases.

	Survival probabilities	
	1 h	48 h
SBC fish survival		
Relative to tailrace control	0.967 (0.940–0.994)	0.958 (0.924–0.988)
Relative to spillbay 2	0.984 (0.945–1.0)	0.983 (0.941–1.0)
Spillbay 2 fish survival		
Relative to tailrace control	0.983 (0.964–1.003)	0.975 (0.952–0.998)

In addition to fish which exhibited visible injuries and died, four noninjured fish (two each from the SBC and Spillbay 2 groups) also died within 48 h (Table 3). None of these fish showed obvious internal injuries when necropsied. However, the two SBC and one Spillbay 2 fish exhibited loss of equilibrium. Two control fish exhibited loss of equilibrium immediately at recapture but were alive and swimming normally at 48 h.

Discussion

Although comparable fish survival probabilities in passage through an SBC are not available in the literature, spillway and bypass survival of juvenile salmonids at several other sites provide a perspective on our estimates (range 0.958–0.983). However, it should be noted that our estimates represent only direct effects of passage while at some other sites estimates may include both the direct as well as indirect passage effects. Depending upon the site-specific characteristics estimated, survival probabilities have generally been less than 1.0 and no passage route appears to be 100% safe. Schoeneman et al. (1961) reported spillway passage survival probability of 0.98 for chinook salmon smolts at McNary Dam on the Columbia River and Big Cliff Dam on the Santiam River. The estimated spill survival probability of chinook fingerlings at Lower Monumental Dam on the Snake River ranged from 0.83 to 0.84 (Long et al. 1972). Heinle and Olson (1981) reported a survival probability of 0.996 for coho salmon *Oncorhynchus kisutch* in passage over the spillway at Rocky Reach Dam on the Columbia River. Ledgerwood et al. (1990) reported spillbay passage survival of juvenile chinook salmon at 1.0 at Bonneville Dam.

Table 3. Injury categories on recaptured chinook salmon smolts released through the surface bypass collector (SBC), Spillbay 2, and in the tailrace (control) at Lower Granite Dam, April 1996. All control fish were alive at 48 h; SBC and Spillbay 2 injured fish died within 48 h.

			Visible injury type		
	Number afflicted	Loss of equilibrium	Bulging eyes	Hemorrhaged eye or head	Hemorrhaged liver or kidney
SBC	4	2		1[a]	1
Spillbay 2	1	1			1[b]
Control	3	2	1		

[a] Small flap of skin peeled on back of head.
[b] Body also was bent.

They also reported that survival was less (up to 14%) for chinook smolts released into a bypass facility than for fish passed through a turbine. All these studies involved a tag–recapture process (e.g., freeze branding, coded wire tags) that occurred over several days and long distances and were not designed to separate direct and indirect effects of spillway passage. However, one recent study by Mathur et al. (1997) at Bonneville Dam estimated survival probability of chinook salmon at 1.0 in passage through spillbays (direct effects). The survival probability of Atlantic salmon smolts *Salmo salar* in passage through ice-log sluices (direct effects) at two hydro dams on the Connecticut River was reported at 0.96 (Heisey et al. 1996). Normandeau Associates (1995) reported a survival probability of about 0.95 for Atlantic salmon smolts in passage through a newly constructed "fish tube" at Vernon Dam on the Connecticut River. We established the overall survival probability of 0.958 in passage through both the SBC and spillway, 0.983 in passage through the SBC alone, and 0.975 through the spillway independent of the SBC. Thus, fish utilizing the SBC are exposed to dual risks of injury or mortality, first through the SBC and second through the spillbay; no additional risk was noted.

The principal causal mechanisms for injury or mortality to fishes transported via spillways are turbulence, shear, pressure changes, variable terminal velocity, potential impact collisions with rock outcrops, abrasive surfaces, obstructions in the flow path, or contact with undersides of bottom opening tainter gates. Therefore, fish survivability may vary depending upon the magnitude of the influence of these factors at each site (Ruggles and Murray 1983). However, the relative importance of factors affecting fish condition and mortality in passage through SBC does not exist in literature. Certain hydraulic factors suspected to cause fish injury or mortality in passage through spillbays include impact velocities against water surface, pressure changes, shear, and turbulence (Bell et al. 1972). Velocities in excess of 16 m/s have been reported to initiate mortality when fish strike standing water while a 6 m/s strike against a solid object can be detrimental (Bell et al. 1972). At Lower Granite Dam, estimated velocities from spill discharge into the stilling basin can exceed 16 m/s. It is unknown whether entrained fish strike solid objects in the flow path. However, some observed injuries appear probably related to hydraulic features. Internal injuries could have resulted from hydraulic forces (pressure reduction) while external injuries may have resulted from contacting solid objects. Since a relatively small number of fish were injured, it is unknown which of the various components in the spillbay and the SBC may contribute to most injuries or mortality.

In summary, while the new concept of SBC may be effective in diverting juvenile salmonids away from turbines, it is not completely benign. In its present configuration it has the potential to inflict some direct injury or mortality on fish. The overall passage survival, as measured by direct effects in the present study, appears to be less than the survival observed at many conventional spillways. To achieve the desired goal of overall enhancement, both fish usage and survival need to be maximized. At present, none of the fish passage routes appear 100% safe at most hydro projects.

References

Bell, M. C., A. C. DeLacy, and H. D. Copp. 1972. A compendium on the survival of fish passing through spillways and conduits. Report prepared for U.S. Army Corps of Engineers Division, North Pacific Corps of Engineers, Portland, Oregon.

Heinle, D. R., and F. W. Olson. 1981. Survival of juvenile coho salmon passing through the spillway at Rocky Reach Dam. Report prepared for Chelan County Public Utility District No. 1, Wenatchee, Washington.

Heisey, P. G., D. Mathur, and T. Rineer. 1992. A reliable tag–recapture technique for estimating turbine passage survival: application to young-of-the-year American shad (*Alosa sapidissima*). Canadian Journal of Fisheries and Aquatic Sciences 49:1826–1834.

Heisey, P. G., D. Mathur, and E. T. Euston. 1996. Passing fish safely: a closer look at turbines vs. spillway survival. Hydro Review 15(4):42–50.

Johnson, G. E. 1996. Fisheries research on the phenomenon in the forebay of Wells Dam in spring 1995 related to the surface smolt bypass. Report prepared for the U.S. Army Corps of Engineers, Walla Walla District, Walla Walla, Washington.

Ledgerwood, R. D., and five coauthors. 1990. Relative survival of sub-yearling chinook salmon which have passed Bonneville Dam via the spillway or the second powerhouse turbines or bypass system in 1989, with comparisons to 1987 and 1988. Report prepared for U.S. Army Corps of Engineers, Portland, Oregon.

Long, C. W., W. M. Marquette, and F. J. Ossiander. 1972. Survival of fingerlings passing through a perforated bulkhead and modified spillway at Lower Monumental Dam, April through May 1972. Progress report to the National Marine Fisheries Service, Seattle, Washington.

Mathur, D., P. G. Heisey, E. T. Euston, J. R. Skalski, and S. Hays. 1996. Turbine passage survival estimation for chinook salmon smolts (*Oncorhynchus tshawytscha*) at a large dam on the Columbia River. Canadian Journal of Fisheries and Aquatic Sciences 53:542–549.

Mathur, D., P. G. Heisey, J. R. Skalski, and M. R. Smith. 1997. Structural modifications at hydro dams: an opportunity for fish enhancement. Pages 358–363 *in* F. M. Holly, Jr., and A. Alsaffar, editors. Energy and water: sustainable development. American Society of Civil Engineers, New York.

Normandeau Associates. 1995. The Vernon bypass fish tube: evaluation of injuries and survival of Atlantic salmon smolts. Report prepared for New England Power Company, Westborough, Massachusetts.

Raemhild, G. A., R. Nason, and S. Hays. 1985. Hydroacoustic studies of downstream migrating salmonids at hydropower dams: two case studies. Pages 244–251 *in* F. W. Olson, R. G. White, and R. H. Hamre, editors. Symposium on small hydropower and fisheries. American Fisheries Society, Western Division and Bioengineering Section, Bethesda, Maryland.

Ruggles, C. P., and D. G. Murray. 1983. A review of fish response to spillways. Canadian Technical Report of Fisheries and Aquatic Sciences 1172.

Schoeneman, D. E., R. T. Pressey, and C. O. Junge, Jr. 1961. Mortalities of downstream migrant salmon at McNary Dam. Transactions of the American Fisheries Society 90:58–72.

Tom K. Iverson, Julie E. Keister, and Robert D. McDonald

Summary of the Evaluation of Fish Passage through Three Surface Spill Gate Designs at Rock Island Dam in 1996

9

In 1996, two designs of surface spill gates were chosen by Chelan County Public Utility District No. 1 (Chelan PUD) for an evaluation of fish passage efficiency of spill at Rock Island Dam. The two gate designs consisted of a 3.4 m wide by 4.3 m deep notch gate and a 9.1 m wide by 2.1 m deep overflow gate. Split-beam hydroacoustic methods were used to estimate the zone of influence of the surface spill gates, and single-beam hydroacoustic methods were used to estimate the number of fish passing the project through each spillbay and powerhouse. Hydroacoustic estimates indicated that notched gates passed significantly more fish per flow than overflow spill gates during spring and summer. Average fish passage per unit flow was significantly higher for some spillbays and seemed to be affected by the volume spilled through adjacent spillbays. The results from this study prompted Chelan PUD to construct six additional notch spill gates for use during the 1997 spill period.

Chapter 9

Introduction

For nearly two decades, Chelan County Public Utility District No. 1 (Chelan PUD), in Wenatchee, Washington, has been studying various bypass methods for safely passing downstream migrating salmonids through Rock Island Dam, while minimizing foregone power production. Hydroacoustic studies have been conducted at Rock Island Dam since 1982. These studies have evaluated various juvenile bypass alternatives including the effectiveness of in-turbine screens as well as the fish passage effectiveness of spillways.

Spill studies at Rock Island Dam have specifically determined that shallow spill (<6.7 m) is more effective than deep spill (>6.7 m) at passing juvenile salmonids through the spillway (Steig and Ransom 1991). Hydroacoustic studies have shown that during both the spring and summer, shallow spill can be nearly 10 times as effective as deep spill in passing downstream migrants at low flow levels (Ransom and Steig 1995).

Because previous spill studies at Rock Island Dam and elsewhere on the Columbia River indicate that surface spill is more effective than deep spill, Chelan PUD, the state and federal fisheries agencies and Treaty Tribes agreed to evaluate the fish passage effectiveness of various surface flow spill gate designs for use as a fish bypass method at Rock Island Dam. Iverson and Steig (1995) showed that most downstream migrating fish were concentrated at the spillbays at this project. Since surface flow appears to be the most effective means of passing fish, this is the most logical long-term solution for passing out-migrating salmonids at Rock Island Dam.

Rock Island Dam is on the Columbia River at river km 729, 24 km southeast of Wenatchee, Washington. The dam has two separate powerhouses, one on each shore, with a spillway between them (Figure 1). An old, submerged road effectively splits the spillway and dam in half at the fish ladder located in the center of the dam. Powerhouse No. 1 is situated

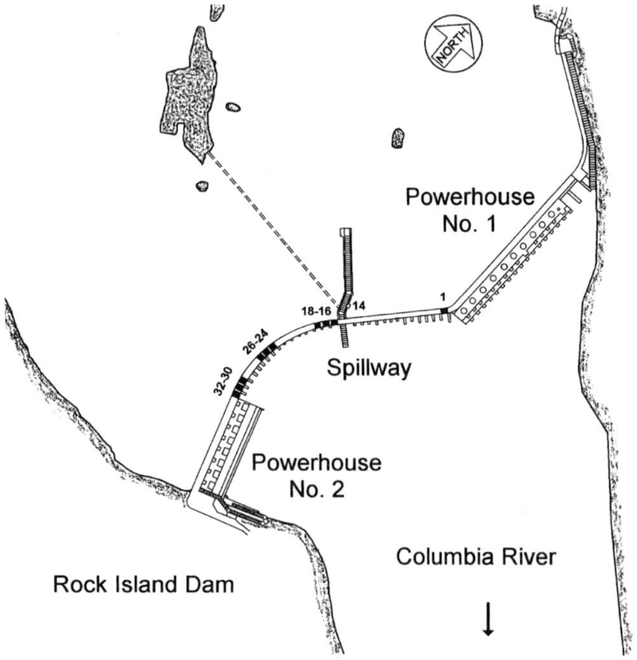

Figure 1. Plan view showing the orientation of Rock Island Dam on the Columbia River. Test spillbays are designated with black boxes.

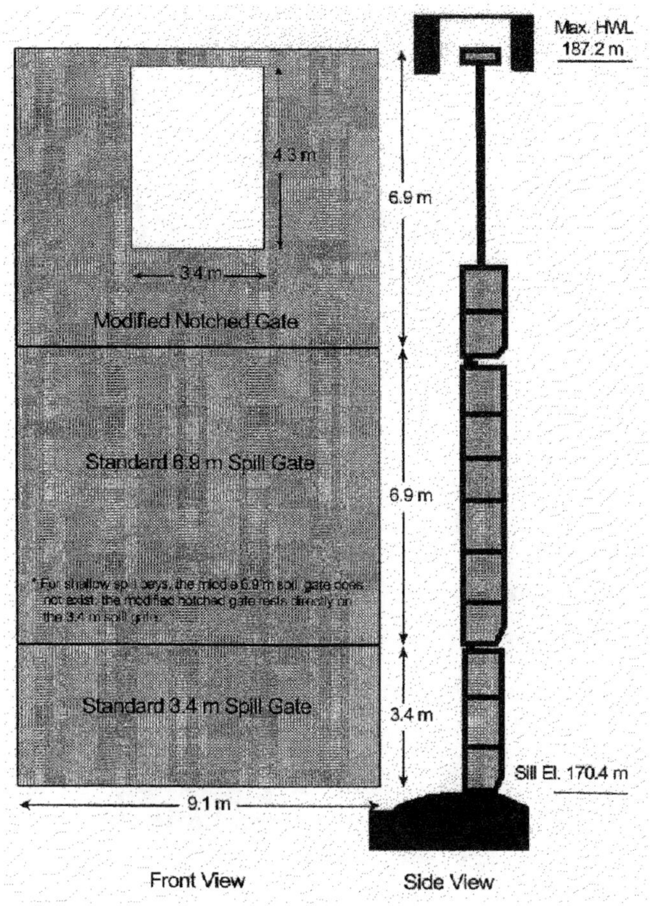

Figure 2. Diagram of the notched spill gate used at Rock Island Dam in 1996.

on the north shore of the river. It is over 213 m long and contains 10 generating units numbered from north to south: four vertical-axis Nagler turbines, and six vertical-axis Kaplan turbine units. Each turbine has three intake galleries 4.6 m wide and 7.6 m high at the head gate slot.

The spillway at Rock Island Dam consists of 31 spillbays, numbered from north to south, and is effectively split by the center fish ladder. There are 14 spill gates on the north side of the center fish ladder and 17 spillbays south of the center fish ladder. The spill gates are 9.1 m wide and are either 10.1 or 16.8 m deep.

Powerhouse No. 2 is situated on the south shore of the river. Powerhouse No. 2 contains eight horizontal-axis bulb turbine units, numbered from north to south. Each turbine has two intake galleries that are 6 m wide by 14.6 m high at the head gate slot. In 1996, two spill gate designs were chosen by Chelan PUD for evaluation of fish passage efficiency of spill in the south spillway of Rock Island Dam: a narrow, deep notched spill gate was to be compared with a wide, shallow overflow spill gate. In addition, an overflow weir spill gate design was to be tested in Spillbay 1, adjacent to Powerhouse No. 1.

The notched spill gate design had an opening of 3.4 m wide by 4.3 m deep in both the shallow and deep spillbays at head water level (HWL) 187.2 m (Figure 2). The average hourly flow from the notched gate in the shallow spillbays (16, 18, and 24) was 53.6 m^3/s. The average hourly flow for the notched gate in the deep spillbays (26, 30, and 32) was 52.4 m^3/s.

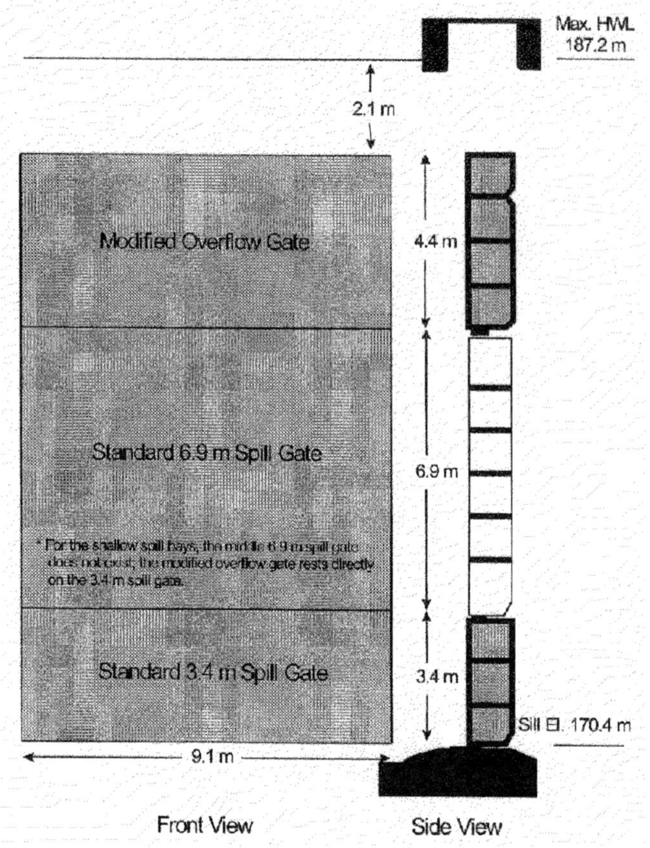

Figure 3. Diagram of the overflow spill gate used at Rock Island Dam in 1996.

In the shallow spillbays the overflow spill gate had an opening of 9.1 m wide by 2.2 m deep at HWL 187.2 m (Figure 3). The average hourly flow in these bays for the overflow gate was 53.2 m^3/s. In the deep spillbays the overflow spill gate had an opening of 9.1 m wide by 2.1 m deep at HWL 187.2 m. The average hourly flow for the overflow gate in these bays was 50.7 m^3/s. The overflow weir in Spillbay 1 at Powerhouse No. 1 had an opening of 9.1 m wide by 3.0 m deep at HWL 187.2 m (Figure 4). The average hourly flow for this weir was 167.2 m^3/s. The standard spill gate used at Powerhouse No. 2 had an opening of 9.1 m wide by 6.5 m deep at HWL 187.2 m. The average hourly flow for this gate was 283.4 m^3/s.

Downstream migrating juvenile salmonids passing Rock Island Dam during the study period included: chinook salmon *Oncorhynchus tshawytscha* (both yearling and sub-yearling), coho salmon *O. kisutch*, sockeye salmon *O. nerka*, and steelhead trout *O. mykiss*.

Methods

Two separate hydroacoustic systems were used to monitor Rock Island Dam in 1996. One system was used to monitor Powerhouse No. 1 and consisted of a 420 kHz echo sounder, a multiplexer/equalizer, a two-channel digital chart recorder, an oscilloscope, and six single-beam transducers. A second system was used to monitor Powerhouse No. 2 and con-

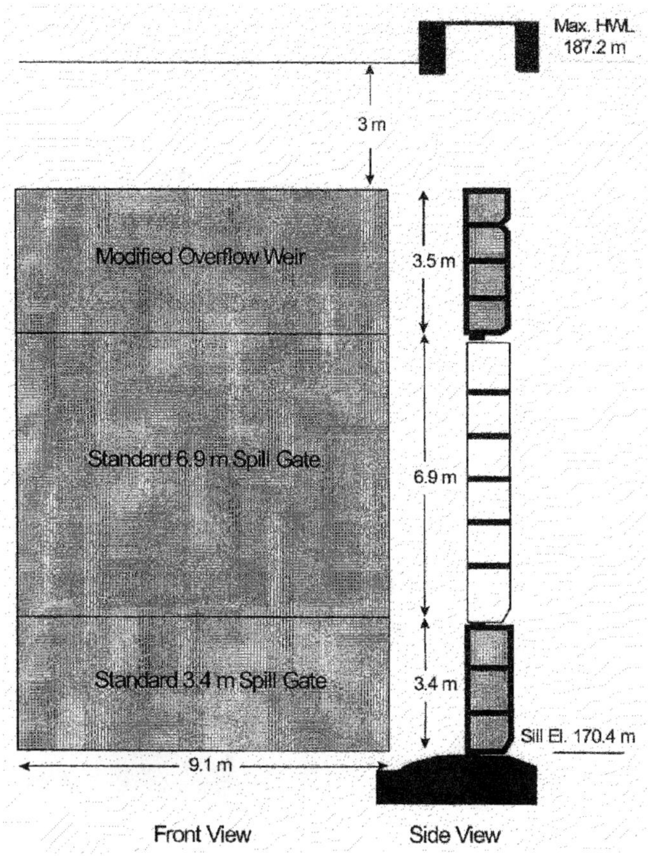

Figure 4. Diagram of the overflow weir used in Spillbay 1 at Rock Island Dam in 1996.

sisted of a 420 kHz echo sounder, a multiplexer/equalizer, 3 thermal chart recorders, an oscilloscope, and 16 single-beam transducers. Both single-beam systems were calibrated and equalized so that only targets of −50 dB and larger would mark on the chart recorder echograms. Thus, targets of −44 dB, the equivalent of a fish 112 mm long (Love 1971) and larger would mark out to the −3 dB points (one-way propagation) of the transducer beam pattern.

The Spillbay 1 transducer was mounted in the center of the sill of the spillbay, 3.7 m upstream of the spill gate. The transducer was aimed straight up then rotated 10° upstream (Figure 5). Under normal operations, this transducer was sampled for 10 min/h. The spillway transducers at Powerhouse No. 2 were mounted below Spillbays 16, 17, 18, 24, 25, 26, 30, 31, and 32. The transducers used in Spillbays 16–18 were 15° circular-beam transducers while the transducers used in Spillbays 24–26 and 30–32 were 6° circular-beam transducers. The spillbay transducers were mounted in the center of the sill of the spillways, 3.7 m upstream of the spill gate. The transducers were aimed straight up then rotated 14–15° upstream, similar to Spillbay 1. During normal operations, these transducers were sampled for a minimum of 3 min/h; however, sample time varied depending on how many monitored spill gates were operating during a sample period.

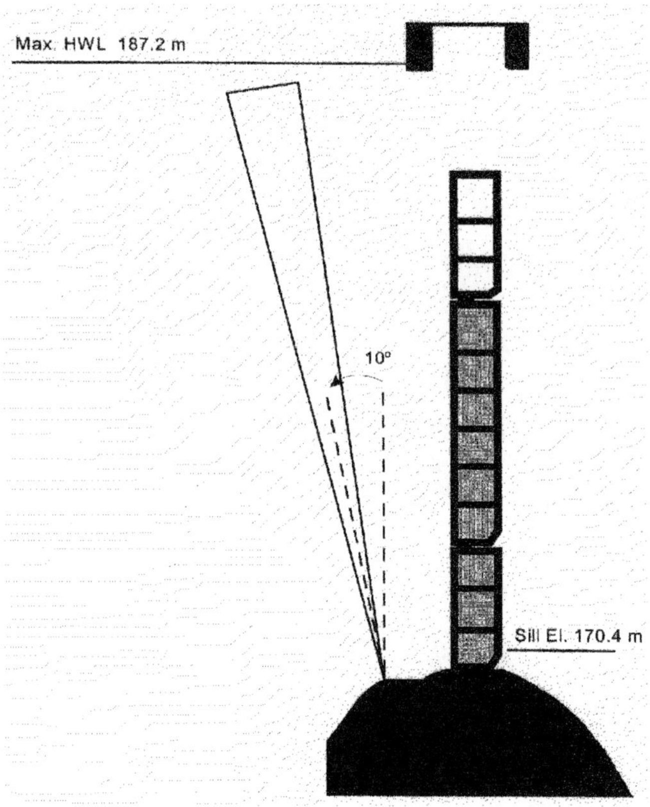

Figure 5. Cross-section of Powerhouse No. 1 spillway showing the mounting location and aiming angle for the transducer at Spillbay 1.

Study Design

To evaluate the optimum surface spill gate design and location of highest fish passage across the spillway, three sets of triplet spill gates were established on either end (Spillbays 16–18 and 30–32) and in the center (Spillbays 24–26) of Powerhouse No. 2 spillway (Figure 1). A spill gate triplet consisted of a standard spill gate in the center with a test gate adjacent to it on either side. Therefore, each triplet would have a standard 6.9 m spill gate in the center and a notched gate on one side with an overflow gate on the other side. The study was organized into randomized 3 d sampling blocks. The test gates within a triplet were randomly assigned to a spillbay on either side of the standard gate every 3 d to eliminate any specific spillbay bias. The standard gate in each triplet was opened for 24 h, 12 h, or kept closed on each day of a sample block to determine the affects of increased flow near the test gates. The gate assignments and daily amount of center spill were randomized and balanced so that each outer gate position experienced an equal number of days for each test gate design. The results of randomization are presented in Tables 1 and 2 and outline the specific gate assignments and operating conditions for the study. Gate assignments in the final block of the study were not random; the final block was used to balance the assignments so that all locations were sampled equally.

Table 1. Spill gate assignments and order of center spill levels for the spring spill study at Rock Island Dam, 1996. V = vertical (notch) gate configuration, H = horizontal (overflow) gate configuration, S = standard gate configuration.

Block	Date	Center spill (h)	Spillbay 16	17	18	24	25	26	30	31	32
1	5/10/96	12	H	S	V	V	S	H	V	S	H
	5/11/96	24	H	S	V	V	S	H	V	S	H
	5/12/96	0	H	S	V	V	S	H	V	S	H
2	5/13/96	12	V	S	H	V	S	H	H	S	V
	5/14/96	24	V	S	H	V	S	H	H	S	V
	5/15/96	0	V	S	H	V	S	H	H	S	V
3	5/16/96	12	V	S	H	H	S	V	H	S	V
	5/17/96	0	V	S	H	H	S	V	H	S	V
	5/18/96	24	V	S	H	H	S	V	H	S	V
4	5/19/96	24	H	S	V	V	S	H	V	S	H
	5/20/96	0	H	S	V	V	S	H	V	S	H
	5/21/96	12	H	S	V	V	S	H	V	S	H
5	5/22/96	12	V	S	H	H	S	V	V	S	H
	5/23/96	0	V	S	H	H	S	V	V	S	H
	5/24/96	24	V	S	H	H	S	V	V	S	H
6	5/25/96	0	H	S	V	H	S	V	H	S	V
	5/26/96	12	H	S	V	H	S	V	H	S	V
	5/27/96	24	H	S	V	H	S	V	H	S	V

The spill study was conducted as a 2 × 3 factorial design with test-gate configuration and hours of center spill as the two factors of interest. Results were analyzed by randomized-block analysis of variance (ANOVAs). Data were blocked by 3 d time periods and location of test gate triplets in the spill way. The randomized-block ANOVA models used were:

fish/flow = block site configuration center_spill site*configuration
site*center_spill configuration*center_spill
site*configuration*center_spill

where: block = 3 d time period, site = location of the test gate in the dam, configuration = vertical (notched) or horizontal (overflow) test gate, center_spill = hours of center standard gate spill (0, 12, or 24 h).

Site, and interactions of configuration and center spill with site, were included to examine relative efficiency of fish passage among spillways. Posthoc tests on hours of center spill and site were calculated by Tukey's honestly significantly different. Null hypotheses were rejected at the α = 0.05 level.

Spring and summer study data were analyzed separately; all data were also analyzed together to allow examination of general trends. Six blocks of data were collected during the spring; one of these blocks was discarded (19–21 May) because center spill deviated from protocol and remained at 24 h spill; there were no 0 or 12 h treatments for that block. Discarding one block of data resulted in unbalanced positioning of the test gates. Vertical (notch) gates occupied Spillbays 18, 24, and 30 twice

Table 2. Spill gate assignments and order of center spill levels for the summer spill study at Rock Island Dam, 1996. V = vertical (notch) gate configuration, H = horizontal (overflow) gate configuration, S = standard gate configuration.

		Center spill	Spillbay								
Block	Date	(h)	16	17	18	24	25	26	30	31	32
1	6/24/96	12	H	S	V	V	S	H	V	S	H
	6/25/96	0	H	S	V	V	S	H	V	S	H
	6/26/96	24	H	S	V	V	S	H	V	S	H
2	6/27/96	0	V	S	H	V	S	H	H	S	V
	6/28/96	24	V	S	H	V	S	H	H	S	V
	6/29/96	12	V	S	H	V	S	H	H	S	V
3	6/30/96	0	H	S	V	H	S	V	H	S	V
	7/1/96	12	H	S	V	H	S	V	H	S	V
	7/2/96	24	H	S	V	H	S	V	H	S	V
4	7/3/96	12	V	S	H	H	S	V	H	S	V
	7/4/96	24	V	S	H	H	S	V	H	S	V
	7/5/96	0	V	S	H	H	S	V	H	S	V
5	7/6/96	0	V	S	H	V	S	H	V	S	H
	7/7/96	24	V	S	H	V	S	H	V	S	H
	7/8/96	12	V	S	H	V	S	H	V	S	H
6	7/9/96	24	H	S	V	H	S	V	V	S	H
	7/10/96	0	H	S	V	H	S	V	V	S	H
	7/11/96	12	H	S	V	H	S	V	V	S	H
7	7/12/96	0	H	S	V	V	S	H	H	S	V
	7/13/96	24	H	S	V	V	S	H	H	S	V
	7/14/96	12	H	S	V	V	S	H	H	S	V
8	7/15/96	0	V	S	H	H	S	V	V	S	H
	7/16/96	24	V	S	H	H	S	V	V	S	H
	7/17/96	12	V	S	H	H	S	V	V	S	H

during the spring study; horizontal (overflow) gates were in those sites three times. The effect of unbalanced treatments within sites was tested by discarding the 16–18 May block, thereby balancing the treatments. Analysis of variance results on balanced, 4-block, spring data were similar to unbalanced 5-block results; therefore, results of the 5-block tests will be discussed. Eight blocks of data were collected during the summer study; test gate configurations were balanced at each site.

Data for ANOVAs were $\log_{10}$ transformed to meet the assumption of homogeneity of variances. Results are presented as untransformed mean ± 1 SE except as noted.

Each fish detection was range weighted to compensate for beam spreading and expanded to the width of the intake being sampled (Iverson 1999, this volume). The expansion factors used for the test spill gates were determined using split-beam hydroacoustics. Fish counts were also expanded for time to determine the hourly fish passage for each intake according to 3–10 min samples per hour.

Fish/flow was calculated on a daily basis as the sum, over 24 h, of the fish passage through one gate divided by the sum, over 24 h, of the hourly average flow through that gate. Hourly data were discarded on a per-hour basis for all hours that gate configuration or center spill was off protocol. Data for days with discarded hours were adjusted to 24 h by dividing the total fish passage and total flow for each gate by the number of hours of data used for the day, then multiplying by 24. For example, days with gate changes in the morning (the beginning of each 3 d block) typically had 1–3 h of time off protocol when gates of the prespecified configuration were not yet in the proper spillways. Those off-protocol hours were discarded and the remaining 21–23 h were normalized to a 24 h time period. Fish/flow data were statistically analyzed as described previously.

Results and Discussion

Single-beam hydroacoustic sampling occurred for a total of 57 d. Sampling began on 23 April and continued through 27 May. Due to high flows in the river, the spring spill study did not start until 10 May. The summer spill study occurred from 24 June through 17 July.

Using the single-beam expansion factors as determined by the split-beam hydroacoustic results (Iverson 1999), and confirmed by hydraulic modeling (Weber et al. 1996), fish passage was calculated for the various spill gates during the spring and summer sampling periods. Fish traces were only accepted down to 6 m deep for the notched gates and down to 5 m deep for the overflow gates. Fish counts for all test gates were expanded to the full width of the spillbays (9.1 m).

Table 3. Results of randomized-block ANOVA testing the effects of time period (block), spillage configuration (config), number of hours of center spill (spill), and position of spill gate in the dam (gate) on fish passage per unit flow for the spring and summer spill studies at Rock Island Dam, 1996.

Source	Degrees of freedom	Mean square	F	Significance
		Spring study		
Block	4	.670	23.72	<.001 [a]
Configuration	1	.254	8.98	.004 [a]
Spill	2	.135	4.77	.013 [a]
Gate	5	.697	24.67	<.001 [a]
Config*Spill	2	.045	1.61	.21
Config*Gate	5	.039	1.40	.24
Spill*Gate	10	.038	1.34	.24
Config*Spill*Gate	10	.030	1.06	.41
		Summer study		
Block	7	.154	4.73	<.001 [a]
Configuration	1	.761	23.31	<.001 [a]
Spill	2	.041	1.25	.291
Gate	5	.820	25.13	<.001 [a]
Config*Spill	2	.018	0.56	.58
Config*Gate	5	.045	1.39	.24
Spill*Gate	10	.356	10.91	<.001 [a]
Config*Spill*Gate	10	.037	1.12	.35

[a] Significant ($\alpha = 0.05$).

Table 4. Fish passage per m³/s flow for all test gates during all center spill configurations during the spring and summer spill studies at Rock Island Dam, 1996.

Gate design	Center spill	Spillbay 16	18	24	26	30	32	Average (f/m³/s)
Spring study								
Overflow	0 h	0.77	0.89	0.60	0.41	2.73	1.86	1.19
	12 h	0.64	1.30	0.47	0.88	2.32	1.77	1.22
	24 h	1.41	1.26	1.31	1.08	1.76	1.77	1.41
	Average	0.94	1.15	0.79	0.79	2.27	1.76	1.27
Notch	0 h	2.15	0.62	1.25	0.73	3.05	2.30	1.67
	12 h	2.42	0.46	1.35	1.20	2.51	2.30	1.77
	24 h	2.80	0.61	1.33	1.16	2.91	2.16	1.92
	Average	2.45	0.56	1.31	1.03	2.82	2.25	1.81
Summer study								
Overflow	0 h	1.14	0.82	0.72	0.63	1.65	1.70	1.11
	12 h	2.40	2.18	0.95	0.49	1.36	1.31	1.45
	24 h	3.49	2.06	1.39	1.03	0.76	0.41	1.53
	Average	2.34	1.68	1.02	0.72	1.26	1.14	1.36
Notch	0 h	2.68	0.79	1.47	0.78	2.74	2.38	1.81
	12 h	4.06	1.71	1.59	0.95	2.09	1.00	1.90
	24 h	4.79	1.54	2.58	0.95	1.41	0.82	2.01
	Average	3.84	1.35	1.88	0.90	2.08	1.40	1.91

Results from randomized block ANOVA tests are presented in Table 3. During the spring study, time period (Block), test gate design (Configuration), number of hours of center spill (Spill), and location within the dam (Gate) significantly affected fish passage/flow. During the summer study, time period (Block), test gate design (Configuration), location within the dam (Gate), and the hours of center spill by location interaction (Spill*Gate) were found significant.

Notched gates passed significantly more fish per unit flow than overflow gates during both the spring and summer spill studies (Table 4). In the spring study, notched gates passed 1.81 ± 0.17 fish per 1.0 m³/s (f/m³/s) of water spilled while overflow gates passed 1.27 ± 0.13

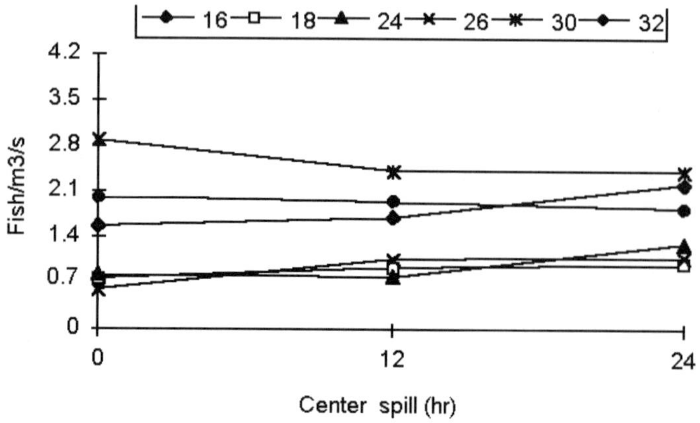

Figure 6. Fish passage per m³/s flow through modified test spill gates as determined by hours of center, standard-gate spill during the spring spill study at Rock Island Dam, 1996.

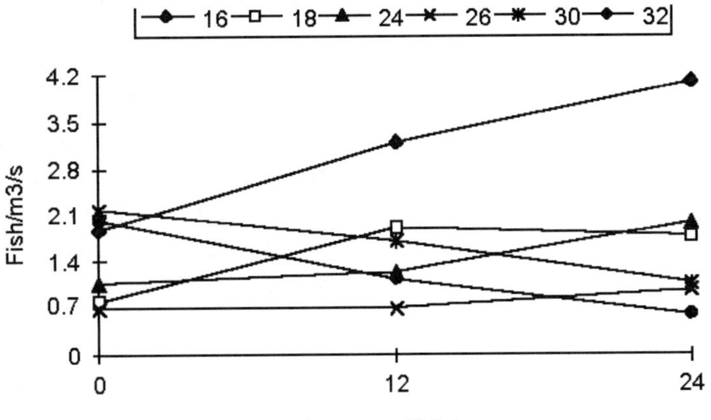

Figure 7. Fish passage per m^3/s flow through modified test spill gates as determined by hours of center, standard-gate spill during the summer spill study at Rock Island Dam, 1996.

f/m^3/s. During the summer, notched gates passed 1.91 ± 1.34 f/m^3/s while overflow gates passed 1.36 ± 0.95 f/m^3/s. This result was consistent across all sites tested on the dam and among all center spill levels.

Number of hours of center spill significantly affected average fish passage per unit flow through the modified gates, but results differed between the spring and summer studies. During the spring study, average number of fish passed per unit flow increased with increasing hours of center spill (Figure 6). There was no difference between 0 h and 12 h treatments or between 12 h and 24 h treatments; however, there was a significant increase between 0 h to 24 h treatments (Tukey's HSD, $\alpha = 0.05$). For the summer study, the effect

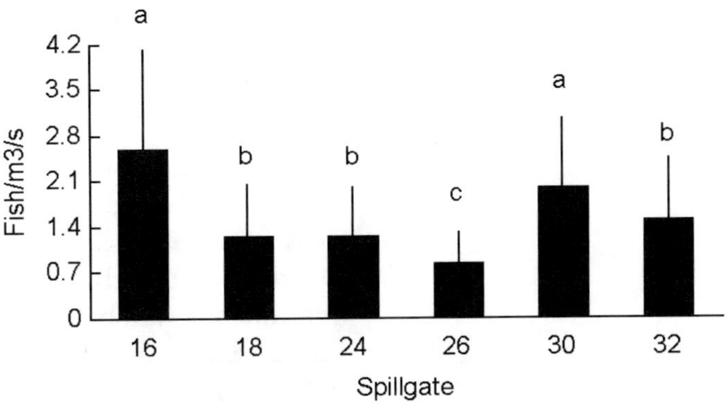

Figure 8. Differences in fish passage per m^3/s flow among spillbays at Rock Island Dam in 1996. Means are averaged across all samples from both the spring and summer spill studies. Means with the same letter are not significantly different by Tukey's HSD. Error bars are 1 SE.

of center spill depended on gate location; fish passage increased with increased hours of center spill at Spillbays 16, 24, and 26, but decreased with center spill at Spillbays 30 and 32 (Figure 7). At Spillbay 18, average fish passage peaked with 12 h of center spill.

Average fish passage was significantly higher per unit flow through some spillbays than through others. For both spring and summer studies combined, Spillbays 16 and 30 had significantly higher fish passage per unit flow than Spillbays 18, 32, 24, and 26 (Figure 8). Spillbay 26 had the lowest average fish passage. However, maximum effectiveness of any particular gate depended on center spill level. In the spring study, the gate*spill level interaction was not significant, but in the summer study, gate effectiveness was affected by the center spill level (Figures 6 and 7).

Spill efficiency of Spillbay 1 was relatively low with 4.9% fish passage compared to 3.1% water spilled during the spring spill study (4.9/3.1 = 1.58 spill index). During the summer spill study, 1.0% fish were spilled at Spillbay 1 through 3.0% of the water (1.0/3.0 = 0.33 spill index). The Powerhouse No. 2 spillbays had a much higher average spill efficiency than Spillbay 1 in both spring (5.6% fish spilled per spillbay and 1.6% water spilled, 5.6/1.6 = 3.50 spill index) and summer (6.5% fish spilled and 1.5% water spilled, 6.5/1.5 = 4.33 spill index).

Conclusions

Notched gates passed significantly more fish per unit flow than overflow gates during both the spring and summer spill studies. In the spring study, notched gates passed 1.81 ± 0.17 f/m^3/s of water spilled while overflow gates passed 1.27 ± 0.13 f/m^3/s. During the summer, notched gates passed 1.91 ± 1.34 f/m^3/s while overflow gates passed 1.36 ± 0.95 f/m^3/s. This result was consistent across all sites tested on the dam and among all center spill levels.

Number of hours of center spill significantly affected average fish passage per unit flow through the modified gates, but results differed between the spring and summer studies. During the spring study, average number of fish passed per unit flow increased with increasing hours of center spill. In the summer study, the effect of center spill depended on gate location. Fish passage increased with increased hours of center spill at Spillbays 16, 24, and 26, but decreased with center spill at Spillbays 30 and 32. At Spillbay 18, average fish passage peaked with 12 h of center spill.

The Powerhouse No. 2 spillbays had much higher average spill efficiency than Spillbay 1 in both spring and summer indicating that the overflow weir was not an effective bypass device. The results of this study prompted the Chelan PUD to construct six new notched spill gates before the 1997 out-migration. The new gates, in addition to the three previously modified notched gates will provide surface spill for fish passage in nine locations through a spill volume of approximately 595.1 m^3/s. Further evaluations will be necessary to determine if this volume of spill results in fish passage rates that are sufficient to meet the Chelan PUD's goals for fish survival.

References

Iverson, T. K., and T. W. Steig. 1995. Hydroacoustic evaluation of the behavior of juvenile salmon and steelhead in the forebay of Rock Island Dam in 1995. Report by Hydroacoustic Technology, Inc. to Chelan County Public Utility District, Wenatchee, Washington.

Iverson, T. K. 1999. A scanning split-beam hydroacoustic technique for determining the zone of entrainment of juvenile salmonids passing hydropower dams. Pages 143-154 *in* M. Odeh, editor. Innovations in fish passage technology. American Fisheries Society, Bethesda, Maryland.

Love, R. H. 1971. Dorsal-aspect target strength of an individual fish. Journal of the Acoustical Society of America 49:816–823.

Ransom, B. H., and T. W. Steig. 1995. Comparison of the effectiveness of surface flow and deep spill for bypassing Pacific salmon smolts at Columbia River basin hydropower dams. Pages 271–280 *in* Waterpower '95: proceedings of the international conference on hydropower. American Society of Civil Engineers, New York.

Steig, T. W., and B. H. Ransom. 1991. Hydroacoustic evaluation of deep and shallow spill as a bypass mechanism for downstream migrating salmon at Rock Island Dam on the Columbia River. Pages 2092–2101 *in* Waterpower '91: proceedings of the international conference on hydropower. American Society of Civil Engineers, New York.

Weber, L., J. DenBleyker, and D. DeJong. 1996. Analysis of Rock Island forebay surface flowlines. Report by the Iowa Institute of Hydraulic Research to Chelan County Public Utility District, Wenatchee, Washington.

Tom K. Iverson

A Scanning Split-Beam Hydroacoustic Technique for Determining the Zone of Entrainment of Juvenile Salmonids Passing Hydropower Dams

10

As surface collection devices and bypass mechanisms become more popular, the intakes that are available for fish passage are becoming smaller and more difficult to monitor using single-beam hydroacoustics. The zone of entrainment into these intakes needs to be estimated to accurately evaluate fish numbers entering the fish bypass and collection devices. In 1996, a vertical scanning split-beam system was used to determine the zone of entrainment for two surface flow spill gate designs. Vertical and horizontal trajectories were determined for fish approaching the two spill gate designs at seven different transducer aiming angles. The width and depth of entrainment were then determined for each test gate design. Based on the split-beam results, expansion factors were developed that could be applied to single-beam hydroacoustic data collected in front of other spillbays where the test gate designs were installed. The scanning split-beam technique clearly defined the zone of entrainment for the two spill gate designs and could be used for other nonstandard intakes at hydropower dams.

Introduction

Runs of Pacific salmon and steelhead trout *Oncorhynchus* spp. on the Columbia and Snake Rivers and their tributaries have been declining since the 1950s. One factor in this decline has been the operation of hydroelectric dams. While most downstream-migrating juvenile salmon and steelhead pass safely through a single dam, the cumulative mortality that results from fish passing through several dams can be substantial (Davidson 1965; Bell et al. 1967; Schwiebert 1977). Over the past two decades, hydropower operators have been studying various bypass methods for safely passing smolts through hydroelectric dams, while minimizing adverse affects on power production.

Since before 1980, hydroacoustic studies have been conducted on the Columbia and Snake Rivers evaluating juvenile salmonid migrations past hydroelectric dams (Thorne and Johnson 1993; Ransom and Steig 1995). Hydroacoustics provide a nonintrusive, nonlethal method for monitoring juvenile fish movement directly in front of turbine intakes and spillbays. At most of these locations, traditional sampling with nets is extremely difficult and usually fatal to the fish being monitored. At Rock Island Dam, a single-beam transducer is typically placed in front of a turbine intake or spillbay, aimed vertically in front of the intake as close as possible to the entrance, and repeatedly sampled within an hour for up to several months (3–10 min/h, 24 h/d, [see Iverson et al. 1999, this volume]). The number of fish counted within each 3–10 min sample at each location is then expanded by the proportion of the hour not sampled to estimate the total number of fish that passed through the intake during a given hour.

$$WF_i = \frac{W_i}{2R_i(\mathrm{Tan}\ 1/2(\theta))}$$

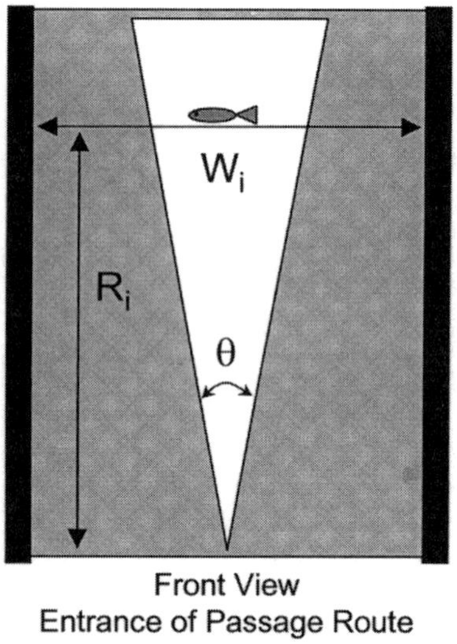

Figure 1. Range-weighting calculation used for single-beam hydroacoustic estimates of fish passage at hydropower dams.

The hydroacoustic count is also expanded for spatial subsampling depending on the range from the transducer at which each fish is observed. Each fish detection is range weighted to compensate for beam spreading, and expanded to the width of the intake being sampled (Figure 1). It is assumed that all fish detected by the single-beam transducer, in the accepted range bins, are entrained to the intake. It is also assumed that all the fish on the same vertical plane as the transducer, across the full width of an intake, are entrained to that intake. Therefore, a fish detected at a specific distance from the transducer is multiplied by the width of the intake and divided by the diameter of the sonar beam at that distance from the transducer to estimate the weighted number of fish entering the intake.

As surface collection devices and bypass weirs become more popular at hydroelectric dams, the intakes that are available for fish passage are becoming smaller and more difficult to monitor using single-beam hydroacoustics. The zone of entrainment into these intakes needs to be determined to accurately evaluate fish numbers entering the surface collectors as compared with other entrances around the dam. The assumption that all fish directly in front of the intake are entrained may not be accurate.

State of the art split-beam hydroacoustic technology has been in use on the Columbia River since 1994 (Steig et al. 1994). Split-beam hydroacoustic techniques allow the user to track fish from ping to ping in three dimensions (Ehrenberg and Torkelson 1996). In this way, direction of travel, velocity, and an accurate target strength can be estimated for each fish.

In 1996, a vertical scanning split-beam system was used to determine the zone of entrainment for two surface flow spill gate designs at Rock Island Dam. The two spill gate designs selected for evaluation of fish passage efficiency consisted of a narrow, deep notched spill gate and a wide, shallow overflow spill gate (Iverson et al. 1996). The general objective of the hydroacoustic study was to use single-beam hydroacoustic technology to compare fish passage through three overflow spill gates with fish passage through three notched spill gates during a nine-week sample period. The zone of entrainment for the single-beam data, and thus the actual expansion widths used for each fish detection, was determined using the vertical scanning split-beam technique described here.

Methods

For a description of Rock Island Dam see Iverson et al. (1999). The notched and overflow spill gates evaluated for this study are also described in Iverson (1999). The test spill gates were placed in Spillbay 26 for the zone of entrainment evaluation.

The hydroacoustic sampling system consisted of a Hydroacoustic Technology, Inc. (HTI) model 243 Digital Split-Beam Echo Sounder (DES), an HTI model 545 Split-Beam Transducer (6° circular, 200 kHz), an HTI model 660 Remote Rotator Controller, a Remote Ocean Systems dual-axis rotator, a computer, a chart recorder, and an oscilloscope. The split-beam hydroacoustic system was calibrated before deployment and data collection.

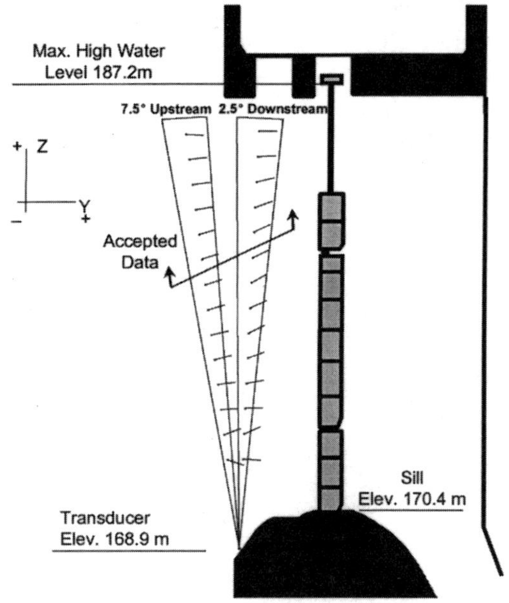

Figure 2. Cross section of the notched test spill gate in Spillbay 26 showing the split-beam transducer mounting location, aiming angles, and the average vertical trajectories.

The system was equalized so that only targets of −53 dB and larger would be accepted as a return from a fish. Thus, targets of −47 dB and larger would mark out to the −3 dB points (one-way propagation) of the transducer beam pattern. This size threshold represents the minimum size of salmonids expected to be passing through the spill gates.

The transducer was installed in the center of the sill of Spillbay 26, 3.7 m upstream of the spill gate, looking directly up in front of Spillbay 26 (Figure 2). Two upstream aiming angles were sampled (7.5° upstream and 2.5° downstream from vertical). The transducer was sequentially aimed at seven different vertical and horizontal positions in front of the spillbay (Figure 3). Each aiming position was sampled for a duration of 4 min, allowing each aim to be sampled twice per hour. The sampling rate was 15 pings per second for all aiming positions. This sampling scheme provided sufficient coverage in front of the spill gates to determine the zone of entrainment for each test spill gate design.

Shadowing due to high fish densities was not an issue at this site since fish densities were low enough that individual target tracking was employed. The aiming angles of the transducer were optimized to ensure that no interaction with the face of the dam was realized. However, shadowing near the surface did occur where the transducer beam hit the top of the spillbay entrance. Therefore, fish were probably not detected in the top 1.5 m of the water column.

Determining zone of entrainment to test spill gates

Since the surface flow test spill gates used at Rock Island Dam were new in design and application, the actual zone of fish entrainment of the test spill gates had to be determined. Single-beam hydroacoustics were used to evaluate the overall passage through all test spill gates. The split-beam

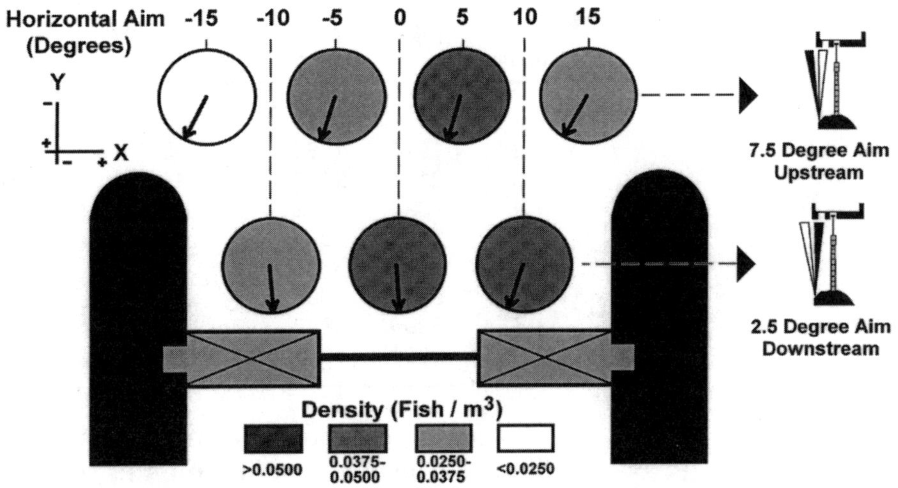

Figure 3. Plan view of the notched test spill gate in Spillbay 26 showing the transducer aiming positions, average horizontal trajectories, and average fish densities.

data were analyzed to determine at what depth and width the fish were entrained to a representative spillbay depending on the test spill gate in place during sampling. From this information, expansion factors were estimated for use with the single-beam hydroacoustic data collected at other spillbays (Iverson et al. 1999). A gate of each test design, notched and overflow, was inserted into Spillbay 26 for two series of three consecutive days. In this way, a total of 6 d of sampling was to occur for each test spill gate design. Sampling occurred 24 h/d with each aiming position sampled for 8 min/h.

Vertical fish trajectories were calculated for each 1 m range bin from the transducer to determine at what depth fish were approaching the test spill gates. The mean fish trajectory for each range bin was calculated for each of the two vertical aiming angles (7.5° upstream and 2.5° downstream). The single beam transducers used for the surface spill study were mounted 0.6 m closer to, and aimed approximately 14–15° away from, the spill gates. Therefore, the single-beam data are more closely represented by the results from the 7.5° upstream vertical aiming angle.

Data analysis

The data collected from the split-beam system consisted of computer data files and printed echograms. The data files were analyzed using a variety of computer programs. The sum channel split-beam data from the DES was printed out as an echogram on a dot matrix printer. These echograms were collected concurrently with the computer data files, and served as a visual reference for ensuring that the data entry technicians were correctly tracking individual fish.

Raw data files were processed using software that presents an electric video echogram that displays each echo received by the echo sounder. Trained technicians were able to track individual fish from ping to ping in three dimensions and enter them into a data file. These data files contain all of the information for each echo for every fish. These data were then summarized into a fish data file to calculate mean trajectory and density

Table 1. Example of fish output data from split-beam analysis software.

Fish number	Start ping	End ping	Total number of echoes	Start X coordinate	Start Y coordinate	Range from transducer/ Start Z coordinate
1	781	823	35	−0.16	0.66	12.94
2	1574	1582	8	0.28	0.33	14.65
3	2131	2169	19	0.68	0.50	13.22
4	2329	2392	50	0.03	0.40	8.15
5	3951	3965	11	0.24	0.34	7.24
6	5312	5318	6	−0.54	0.16	10.34
7	5952	6005	50	−0.12	0.45	8.68
8	6296	6348	53	−0.23	−0.11	8.32
9	6354	6388	30	0.61	0.52	13.16

by aiming position. An example of the fish data summary is presented in Table 1. The fish data file contains fish number, start ping, end ping, total number of echoes, start X coordinate, start Y coordinate, start Z coordinate, range from transducer, distance traveled in X dimension, distance traveled in Y dimension, distance traveled in Z dimension, swimming speed, target strength, and target strength standard deviation for each fish detected. After tracking individual fish, data files were combined for each transducer aiming angle. Data tables summarizing fish trajectory, fish velocity, and fish target strength by range were produced.

The direction of fish movement helped in determining the zone of entrainment for the test spill gates. For each sampling configuration, the average distance traveled in the X, Y, and Z (range) dimensions was calculated for each 1 m range cell for each aiming angle. The vertical trajectory angles were calculated from the slope of the line determined by the distance traveled in the Y and Z dimension for all fish showing a positive movement toward the face of the dam. The average horizontal trajectory angles were calculated from the slope of the line determined by the distance traveled in the X and Y dimensions for all fish showing a positive movement toward the face of the dam. The average horizontal and vertical trajectories were then displayed on an image of the transducer beam to show how the fish moved spatially near the test gates.

Fish density was calculated and displayed for each aiming angle. For each sampling location, fish were grouped into 1 m range cells for each aiming angle. The number of fish detected in each cell was divided by the acoustic sample volume of that range cell to determine the density of fish per unit sample volume (in units of fish/m^3) for each minute sampled. The formulas used to calculate spatial fish density are provided in Steig and Johnston (1995).

The size of each fish affects the effective beam width (sample volume) of the transducer, since larger fish are seen further off the acoustic axis than smaller fish. The split-beam hydroacoustic method estimated an acoustic size for each fish (Ehrenberg 1982), which was then taken into account during postprocessing.

Results and Discussion

Split-beam sampling occurred at Spillbay 26 from 1900 hours on 27 June through 0700 hours on 8 July 1996, 24 h/d. During this period, a notched test gate was installed in Spillbay 26 for approximately 6 d and an overflow test gate was installed for approximately 4.5 d (Table 2). Standard

Table 1. (Continued.)

Distance traveled in X dimension	Distance traveled in Y dimension	Distance traveled in Z dimension	Swimming speed	Target strength	Target strength SD
0.11	−1.37	1.57	1.47	−43.77	3.071
0.04	−0.27	0.06	1.38	−45.81	2.062
−0.11	−1.03	1.24	1.20	−43.52	2.030
0.37	−0.85	0.81	0.64	−41.62	3.556
0.12	−0.19	0.11	0.37	−34.21	2.895
0.11	−0.11	0.12	0.86	−39.76	3.850
0.34	−0.93	0.67	0.74	−39.51	4.166
−0.13	−0.35	0.70	0.69	−42.17	1.760
−0.30	−1.16	0.64	1.44	−41.46	4.430

spill also occurred in Spillbay 25 during this time period in 12 h and 24 h blocks with periods of no spill between blocks. No spill occurred in Spillbay 27 during these tests. To determine the zone of entrainment, all data were combined for each test gate design regardless of Spillbay 25 operations. This condition best represented the other test spill gates being monitored by single-beam hydroacoustics. Although Spillbay 25 operations did have minor affects on the average horizontal trajectories in front of the test spill gates, the final zone of entrainment determination at each level of Spillbay 25 operations was similar.

Fish trajectories approaching the notched test gate

Split-beam data were collected for a total of 106.4 h while the notched test gate was in Spillbay 26. A total of 3,524 fish were tracked during this time period with 3,288 fish (93.3%) observed moving toward the dam. For each vertical aiming angle (7.5° upstream and 2.5° downstream), the approach trajectories were averaged across the width of the spillbay. Vertical trajectories indicated that fish below approximately 6 m, on average, would

Table 2. Date and time of spill gate activity in Spillbays 25 and 26 at Rock Island Dam during split-beam monitoring from 27 Jun through 8 Jul 1996. (Gate 27 was closed for the duration of this study.)

Date	Time	Activity	Gate 26 Configuration	Gate 25 Operations
27 Jun	1900	Begin sampling	Overflow	Closed
28 Jun	0700	Open gate 25	Overflow	Open (24h)
29 Jun	0700	Close gate 25	Overflow	Closed
29 Jun	1900	Open gate 25	Overflow	Open (12h)
30 Jun	0700	Gate change		
30 Jun	0900	Resume sampling	Notched	Closed
1 Jul	1900	Open gate 25	Notched	Open (12h)
2 Jul		Ongoing sampling	Notched	Open (24h)
3 Jul	0700	Close gate 25	Notched	Closed
3 Jul	1900	Open gate 25	Notched	Open (12h)
4 Jul		Ongoing sampling	Notched	Open (24h)
5 Jul	0700	Close gate 25	Notched	Closed
6 Jul	0700	Gate change		
6 Jul	0900	Resume sampling	Overflow	Closed
7 Jul	0700	Open gate 25	Overflow	Open (24h)
8 Jul	0700	Close gate 25	Overflow	Closed

Table 3. Vertical distribution of downstream-migrating fish detected in front of Spillbay 26 at Rock Island Dam using vertical scanning split-beam hydroacoustics from 27 Jun through 8 Jul 1996.

Depth (m)	Notched test gate: Number of fish	Notched test gate: Percent	Notched test gate: Cumulative percent	Overflow test gate: Number of fish	Overflow test gate: Percent	Overflow test gate: Cumulative percent
0–1	210	6.39	6.39	196	7.4	7.4
1–2	337	10.25	16.64	341	12.9	20.2
2–3	455	13.84	30.47	421	15.9	36.1
3–4	406	12.35	42.82	362	13.7	49.8
4–5	384	11.68	54.50	304	11.5	61.2
5–6	326	9.91	64.42	229	8.6	69.9
6–7	290	8.82	73.24	200	7.5	77.4
7–8	225	6.84	80.08	170	6.4	83.8
8–9	230	7.00	87.07	138	5.2	89.0
9–10	122	3.71	90.78	106	4.0	93.0
10–11	123	3.74	94.53	79	3.0	96.0
11–12	83	2.52	97.05	42	1.6	97.6
12–13	47	1.43	98.48	35	1.3	98.9
13–14	35	1.06	99.54	21	0.8	99.7
14–15	9	0.27	99.82	8	0.3	100.0
15–16	6	0.18	100.00	0	0.0	100.0
Sum	3,288			2,652		

not be entering the notched test gate (Figure 2). In addition, over 64% of the fish observed approaching the notched test gate were within the top 6 m of the water column (Table 3). This suggests that only fish in the top 6 m of the single-beam hydroacoustic results should be accepted as entrained fish into the notched test gate.

Horizontal fish trajectories for each aiming angle were calculated by averaging all observed fish trajectories to 6 m in depth. The horizontal trajectories indicated that fish directly in front of the notched gate (2.5° downstream), across the full width of the spillbay, were entrained to the notch (Figure 3). Also, fish observed where the single-beam transducer coverage was located (7.5° upstream) appeared to be entrained across the full width of the spillbay. This suggests that the expansion width for calculating fish passage through the notched gate should be the full 9.1 m width of the spillbay. Figure 3 also presents the average fish densities for each aiming angle. Higher fish densities were observed on the north side of the spill gate, near Spillbay 25, although fish were distributed across the entire width of the intake.

Based on the split-beam results presented above, single-beam data were accepted only within the top 6 m of the water column for the notched test gates. Also, fish counts were expanded to the full width of the spillbay (9.1 m).

Fish trajectories approaching the overflow test gate

Split-beam data were collected for 76.2 h while the overflow test gate was in Spillbay 26. A total of 2,824 fish were tracked during this time period with 2,652 fish (93.9%) moving toward the dam. For each vertical aiming angle (7.5° upstream and 2.5° downstream), the approach trajectories were averaged across the width of the spillbay. Vertical tra-

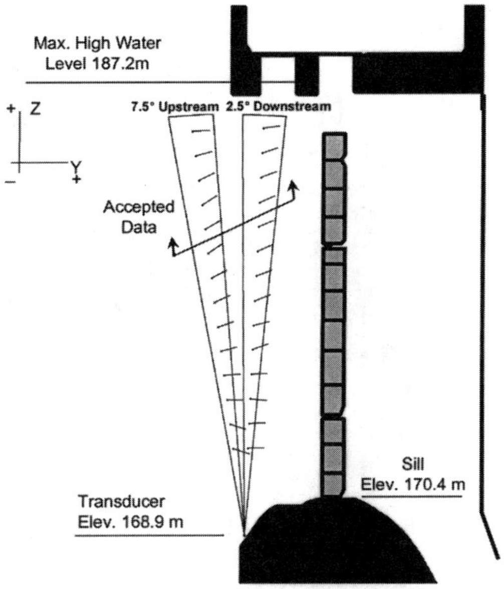

Figure 4. Cross section of the overflow test spill gate in Spillbay 26 showing the split-beam transducer mounting location, aiming angles, and the average vertical trajectories.

jectories indicated that fish below approximately 5 m would not be entering the overflow test gate (Figure 4). In addition, over 61% of the fish observed approaching the overflow test gate were within the top 5 m of the water column (Table 3). This suggests that only fish in the top 5 m of the single beam hydroacoustic results for the overflow test gate should be accepted as entrained fish. Horizontal fish trajectories for each angle were calculated by averaging all observed fish trajectories to 5 m in depth. The horizontal trajectories indicated that fish directly in front of the overflow gate (2.5° downstream), across the full width of the spillbay, were entrained by this test gate (Figure 5). Also, the fish where the single-beam transducer coverage was located (7.5° upstream), appeared to be entrained across the full width of the spillbay. This suggests that the expansion width for calculating fish passage through the overflow gate should be the full 9.1 m width of the spillbay. Figure 5 also presents the average fish densities for each aiming angle. For this test gate, the densities are more evenly spread across the width of the spillbay but highest in the center of the gate.

Based on the split-beam results, single-beam data for the overflow test gates were accepted only within the top 5 m of the water column. Also, the data were expanded to the full width of the spillbay (9.1 m).

Conclusions

The vertical scanning split-beam technique provided an excellent method for determining the zone of entrainment to the two surface spill gate designs. The split-beam technique allowed the absolute direction of travel to be determined for each fish detection. If expansion widths would have been determined solely by the physical width of

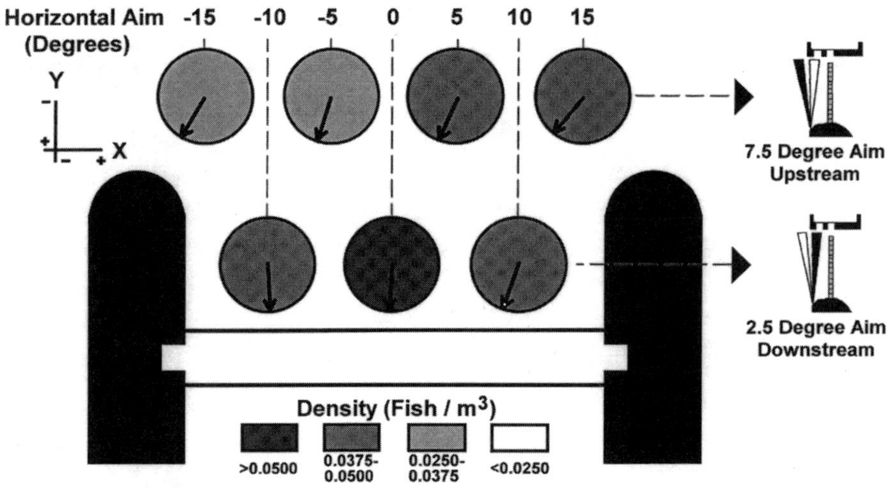

Figure 5. Plan view of the overflow test spill gate in Spillbay 26 showing the transducer aiming positions, average horizontal trajectories, and average fish densities.

each test spill gate, the expansion factor for the notched gate would have been almost one-third the width measured by the split-beam technique (3.4 m instead of 9.1 m). Had this error gone unnoticed, the fish passage estimate for the notched gate would have been biased low by almost a factor of three. The results obtained would have been significantly different and decisions would have been made using incorrect results.

In 1997, this technique was applied to representative passage routes for each powerhouse and spillway at Rock Island Dam to ensure that accurate zone of entrainment calculations were being used for the ongoing single-beam hydroacoustic data being collected. This ensured the best possible fish passage estimates at all sampling locations.

Split-beam hydroacoustics allows the user to determine the three-dimensional trajectories of fish in front of an intake. The scanning system increases the area of coverage for a single transducer and allows the entire zone of entrainment to be mapped. From this technique, the width and depth of fish entering an intake can be determined. These expansion factors can then be applied to single-beam hydroacoustic data at similar intakes across a hydropower project. This technique will assure that single-beam data accurately represents fish passage through nonstandard intakes at hydropower projects.

References

Bell, M. C., A. C. Delacy, and G. J. Paulik. 1967. A compendium on the success of passage of fish through turbines. Prepared for the U.S. Army Corps of Engineers, Portland, Oregon.

Davidson, F. A. 1965. The survival of the downstream migrant salmon at the power dams and in their reservoirs on the Columbia River. Report to Public Utility District of Grant County, Ephrata, Washington.

Ehrenberg, J. E. 1982. A review of in situ target strength estimation techniques. FAO (Food and Agriculture Organization of the United Nations) Fisheries Report 300:85–90.

Ehrenberg, J. E., and T. C. Torkelson. 1996. Application of dual-beam and split-beam target tracking in fisheries acoustics. ICES Journal of Marine Science 53:329–344.

Iverson, T. K., J. E. Keister, and T. W. Steig. 1996. Hydroacoustic evaluation of three surface flow spill gate designs and overall fish passage at Rock Island Dam in 1996. Report by Hydroacoustic Technology, Inc. to Chelan County Public Utility District No. 1, Wenatchee, Washington.

Iverson, T. K., J. E. Keister, and R. D. McDonald. 1999. Summary of the evaluation of fish passage through three surface spill gate designs at Rock Island Dam in 1996. Pages 129-142 *in* M. Odeh, editor. Innovations in fish passage technology. American Fisheries Society, Bethesda, Maryland.

Ransom, B. H., and T. W. Steig. 1995. Comparison of the effectiveness of surface flow and deep spill for bypassing Pacific salmon smolts (*Onchorynchus* spp.) at Columbia River basin hydropower dams. Pages 271–280 *in* Waterpower '95: proceedings of the international conference on hydropower. American Society of Civil Engineers, New York.

Schwiebert, E., editor. 1977. Columbia River salmon and steelhead. American Fisheries Society, Special Publication 10, Bethesda, Maryland.

Steig, T. W., R. A. Adeniyi, and T. K. Iverson. 1994. Hydroacoustic evaluation of the behavior of juvenile salmon and steelhead approaching the powerhouse and forebay of Rocky Reach Dam during 1994. Report by Hydroacoustic Technology, Inc. to Chelan County Public Utility District No. 1, Wenatchee, Washington.

Steig, T. W., and S. V. Johnston. 1995. Monitoring fish movement patterns in a reservoir using horizontally scanning split-beam techniques. ICES Journal of Marine Science 53:435–441.

Thorne, R. E., and G. E. Johnson. 1993. A review of hydroacoustic studies for estimation of salmonid downriver migration past hydroelectric facilities on the Columbia and Snake rivers in the 1980s. Reviews in Fisheries Science 1(1):27–56.

John Hedgepeth, David Fuhriman, and William Acker

Fish Behavior Measured by a Tracking Radar-Type Acoustic Transducer Near Hydroelectric Dams

11

Recent studies of fish behavior around hydroelectric dams have used acoustics with split-beam methodology. A complementary methodology called the tracking transducer takes advantage of split-beam capabilities for expanding fish behavior investigations.

In a fixed deployment, the transducer remains static while fish move through the beam. Transducers mounted on rotators allow expansion of spatial coverage, but tracking is still limited for individual fish. In this study we applied the principle of tracking radar, aligning the antenna beam axis with a target, with an acoustic transducer and dual-axis rotators to track individual fish over longer periods of time. Deviation of the target from the beam axis produces a correction to point the axis toward the target. At the same time, data regarding the fish position and movement and acoustic size are recorded to hard disk. Individual fish tracks are visualized in AutoCAD.

Tracks at Ice Harbor Dam, Snake River in 1995 showed that fish were drawn into the bypass sluiceway when it operated, and that the depth of fish, as they approached the dam, determined turbine entrainment. In 1996, at The Dalles Dam, Columbia River, the tracking transducer showed that fish trajectories were steeper into turbine intakes when occlusion plates were installed in front of the intakes. The study of fish behavior around spillway overflow weirs at John Day Dam, Columbia River, showed that as near-surface fish approached the weirs they sounded and attempted to move away from the spillway. In general, however, most fish tended to follow streamlines of flow, except later in the season when nonsalmonid species were present.

The tracking split-beam offers the possibility of providing intermediate track lengths and detailed behavioral information of individual fish in the near-dam forebay hydraulic environment. This should be another valuable tool to assist in evaluation of fish bypass or collection facilities.

Introduction

Since the early 1980s acoustical systems have been applied to study the behavior of fish near and around hydroelectric power plants (Thorne and Johnson 1993). In most cases, a transducer or a series of transducers are deployed in fixed positions and aimed at a predetermined angle to maximize the transducer beam pattern within a defined area of concern.

In the usual fixed deployment, the transducer remains in a static position while fish move through the beam. The static position of the transducer limits the amount of information describing the direction of movement and behavior of fish.

The principle of tracking radar, aligning the antenna beam axis with a target, has been applied with an acoustic transducer for tracking fish targets. Little civilian application of tracking radar principles has been applied in hydroacoustics, although monopulse (meaning an estimate per transmission pulse) or tracking radars have been widely used for tracking aircraft. Previous acoustic work includes a phase-steered array composed of many elements, which is very complicated and expensive (Jaffe 1995). Our approach was relatively simple: control of the aim of the transducer is accomplished with a personal computer and enclosed (waterproof) stepper motors. The new result is called the tracking transducer.

Monopulse radar principles have been well established (Sherman 1984). Deviation of the target from the beam axis produces a correction that is applied to drive the target toward the axis. The earliest tracking radars were conical scanners. Present day tracking radars use monopulse or simultaneous lobing. By definition, monopulse systems estimate target parameters with a single pulse. A split-beam transducer is an example of an acoustic monopulse device.

A split-beam transducer was attached to a dual-axis rotator system to track fish targets in the same way that monopulse radars track aircraft. Tracking sonar is useful in determining direction of fish travel. Moreover, the mechanical system allows a considerable expansion in traditional sampling volume. Data from tracking sonar are more easily quantifiable than sector scanning sonar results because of our understanding of scientific quality split-beam echosounders.

The fish behavior studies determined direction, velocity, path, and vertical and horizontal placement of fish within the water column in front of turbines and spillways. Fish behavior was determined by actively tracking fish with an acoustic transducer and recording their positions and acoustic sizes over the time period they could be "locked on to." This development is an extension of split-beam methodology (MacLennan and Simmonds 1992).

The first application of the tracking sonar, at Ice Harbor Dam on the Snake River (Figure 1), demonstrated the advantage of employing motor-controlled transducers in acoustic surveys of fish behavior in the vicinity of fish bypass systems (BioSonics 1996). Building upon that experience, the system was deployed at The Dalles Dam on the Columbia River near turbine intake occlusion structures (BioSonics 1997a). Two tracking systems were recently used to determine fish reactions in response to installation of spillway overflow weirs at John Day Dam. In addition, acoustic tags have been successfully tracked, holding promise for other future uses in determining fish behavior (Johnson et al. 1998).

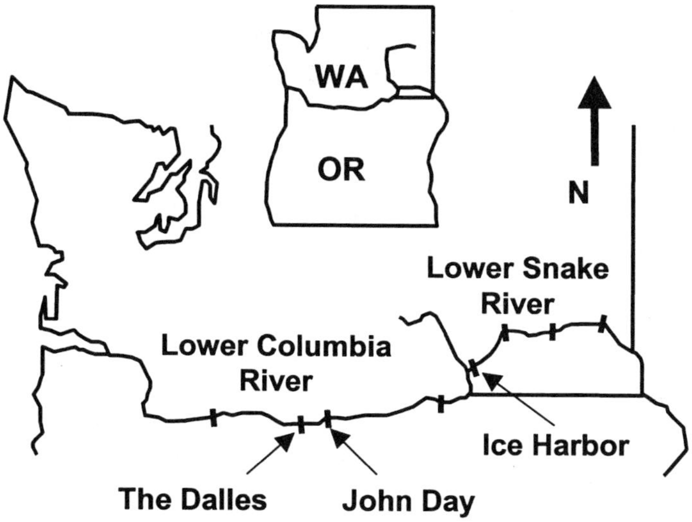

Figure 1. Location of Ice Harbor Dam on the Snake River, The Dalles Dam, and John Day Dam on the Columbia River.

Methods

The output of the split-beam echosounder provides information on fish location. In general, a split-beam transducer is electrically divided into two orthogonal sets of paired receivers. An acoustic signal is transmitted and reflected from a fish. This echo encounters the two sets of receiving elements, allowing the direction of arrival of an echo to be determined. An acoustic wave front propagating towards the transducer arrives at different times at the four quadrants, causing the phase angle of the electrical output signal from the receivers to differ. One angle (often called alongships) is determined from the electrical phase difference between one set of receiving elements, and a second angle (athwartships) is estimated from the orthogonal elements (MacLennan and Simmonds 1992).

These phase angles in the form of a telegram can be outputted via a serial port of an acquisition PC (which is connected to the split-beam echosounder) to a second PC. The telegram contains single-echo detections for one ping: header, time tag, number of single echo detections, depth (meters), compensated target strength (the estimated acoustic size of the fish, referred to as TS in dB), alongships angle (degrees), and athwartships angle (degrees). The second PC containing tracking control software captures the telegram. The transducer is installed in a two-axis aiming armature rotated by high-speed stepper motors that are controlled by the second computer receiving the echosounder directional signals. The stepper motor control software receives the alongships and athwartships angle measurements and then actuates the stepper motors to keep the main axis of the transducer beam aimed on the target, thereby tracking the target (Figure 2). Software had the capability to "lock on to" fish targets and "radar track" their trajectories while continually changing the transducer angle to intercept the fish.

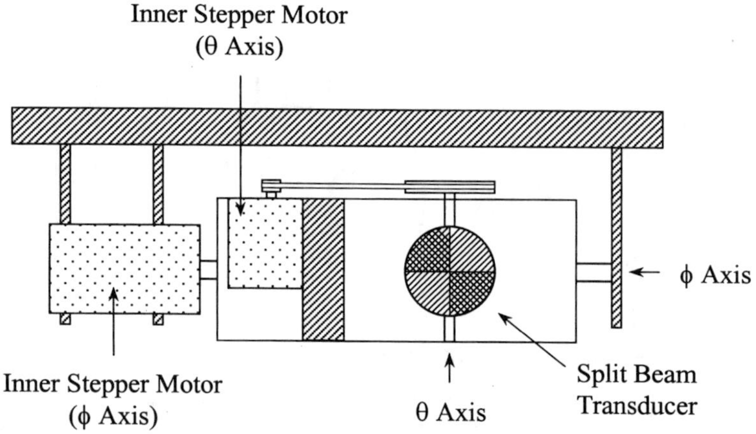

Figure 2. Diagram of the armature of the Tracking Transducer System. The angles θ and γ are used with two split-beam angles, ϕ and ψ, to calculate new motor angles to follow a moving fish target.

The operation of the tracking transducer is simple: center the fish on the acoustic axis and then follow its path while measuring its acoustic size. To accomplish this, the split-beam phase angles, γ (alongship) and ψ (athwartship) needed to be compensated by the stepper motor angles θ and ϕ (Figure 2).

The following mathematical analysis assumes an absolute coordinate system (x, y, z). The x-axis runs along the dam to the right when facing away from the dam, the y-axis points up, and the z-axis points away from the dam. Let the unit vectors in the absolute coordinate system be $\hat{i}$, $\hat{j}$, and $\hat{k}$. Then unit vectors of the rotated coordinate system (ξ, η, and ζ) of the transducer are:

$$\begin{aligned} e_\xi &= \cos\theta\hat{i} - \sin\theta\sin\phi\hat{j} - \sin\theta\cos\phi\hat{k} \\ e_\eta &= \cos\phi\hat{j} - \sin\phi\hat{k} \\ e_\zeta &= \sin\theta\hat{i} - \cos\theta\sin\phi\hat{j} + \cos\theta\cos\phi\hat{k} \end{aligned} \tag{1}$$

The unit vector to a fish target is approximately:

$$e_\rho = \sin\psi e_\xi - \sin\gamma e_\eta - \sqrt{1-\sin^2\gamma - \sin^2\psi e_\zeta} \tag{2}$$

In terms of the stepper motor coordinate system, the unit vector to the fish is:

$$\begin{aligned} e_\rho = &\left(\cos\theta\sin\psi + \sin\theta\sqrt{1-\sin^2\gamma-\sin^2\psi}\right)\hat{i} \\ &+\left(-\sin\theta\sin\phi\sin\psi + \cos\phi\sin\gamma + \cos\theta\sin\phi\sqrt{1-\sin^2\gamma-\sin^2\psi}\right)\hat{j} \\ &+\left(-\sin\theta\cos\phi\sin\gamma - \sin\phi\sin\gamma + \cos\theta\cos\phi\sqrt{1-\sin^2\gamma-\sin^2\psi}\right)\hat{k} \end{aligned} \tag{3}$$

The new stepper motor angles required to follow the fish are:

$$\theta' = \sin^{-1}\left(\cos\theta \sin\psi + \sin\theta\sqrt{1-\sin^2\gamma - \sin^2\psi}\right)$$

$$\phi' = \tan^{-1}\left(\frac{-\sin\theta \sin\phi \sin\psi + \cos\phi \sin\gamma + \cos\theta \sin\phi\sqrt{1-\sin^2\gamma-\sin^2\psi}}{-\sin\theta \cos\phi \sin\psi - \sin\phi \sin\gamma + \cos\theta \cos\phi\sqrt{1-\sin^2\gamma-\sin^2\psi}}\right) \quad (4)$$

Ice Harbor Dam, spring and summer 1995

Much of the methodology was developed for the tracking transducer in the Ice Harbor smolt behavior study. The fish behavior study required the knowledge of direction, velocity, path, and vertical and horizontal placement of fish within the water column in a 15 m radius in front of the sluiceway collectors. The sluiceway slots were manipulated by being either open or closed.

The tracking transducer system consisted of a PC-controlled 120 kHz Simrad EY 500 split-beam fisheries echosounder and split-beam transducer, a second PC with Keithley-Metrabyte dual-axis controller board, interface with stepper motor drivers and limit switch detection hardware, and an underwater armature using two Superior stepper motors.

The tracking transducer was located about 10 m below the surface at turbine unit 4B and at the edge of the 1.8 m vertical slot (Figure 3). This 7° (full beamwidth at half power) circular split-beam transducer was used to monitor behavior in front of the sluiceway and slot at unit 4.

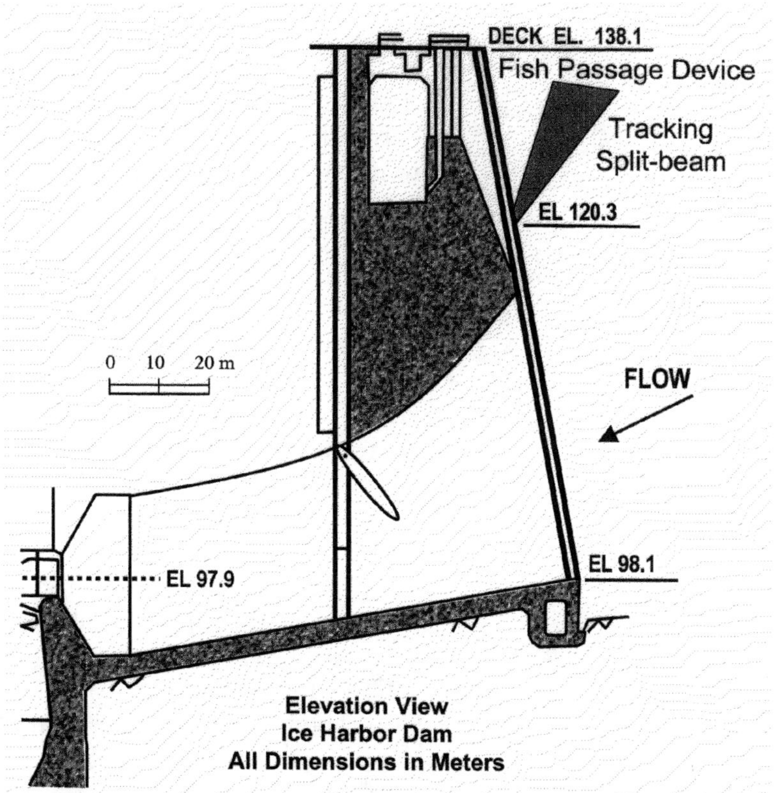

Figure 3. Location of the tracking transducer at Ice Harbor Dam, 10 m below the surface at turbine unit 4B and into the side of a 1.8 m vertical slot for fish passage attraction into a sluiceway.

During operation, the transducer rotator angles were reset every 10 min to point in a specific direction systematically chosen. The echosounder was operated at approximately 10 pings/s. However, because of inability to control pinging by the stepper motor computer and the stepper motor movement, which could compromise accurate position estimation, every other ping was discounted, and as a result the effective operation used 5 pings/s. As the system tracked targets, the transducer moved to point toward the last target position. When a target was lost, the transducer did not move until a new target was acquired, or it was reset at the end of 10 min.

Target selection and tracking was controlled by a number of parameters. The selection parameters include initial target position and returned signal strength. The range of compensated target strength in which the system would select a new target to track was −60 dB to −30 dB. Once selected, the target's signal strength on subsequent pings was ignored. The target range criterion originally was a minimum distance (radius) from the transducer, between 1.0 and 15.0 m. Targets outside this range would not be selected for tracking. To better define the survey volume, the position criterion was changed on 6 June to Cartesian (X, Y, Z) limits, which were: along dam, −20.0–20.0 m; away from dam, −2.0–23.0 m; and up/down, −15.0–10.0 m. Positive values are right (approximately to the south), away (upstream, approximately east), and up, respectively. Once selected, a target was tracked until lost or physical tracking limits were reached.

Tracking parameters were used to define tracking errors. If the target density was too high, the system was unable to tell which target matched the one tracked on the previous ping. If the second-closest target was less than three times as far away as the first choice, the track was ended. If the speed of the tracked target was greater than 2.0 m/s, the track was ended. If a target could be tracked immediately following the end of the previous track, a mark was added to the file to show that the best possible target continued to be tracked. The data were inspected manually at those marks to make the determination if it was a new target. In later studies, at other locations, tracking parameters evolved as the tracking system became better understood.

Other tracking inconsistencies were possible due to irregular system operation or errors. If there was more than 1.0 s between pings, a track was marked as ended. If there were two pings at the same time (due to the limited precision of the EY500 sonar time tags), the track was marked as ended. As with the other tracking errors, these points in the data were examined manually if necessary to determine what occurred. Files were filtered for a minimum of 5 pings per track, and the resulting files were examined manually for tracking errors, date inconsistencies, and system positioning errors.

A combined file was manually divided into separate files based on the time ranges that the sluiceway gate was open or closed. These files were used to generate Initial Graphics Exchange Specification (IGES) files (.IGS extension) for AutoCAD plotting and statistics files (*.STA) extension. The IGES format is a standard CAD drawing exchange format used to transfer drawings between different CAD programs. The three-dimensional *.IGS file data were then read by AutoCAD and examined in various orientations. To mark target direction for monochrome plots, tracks end with a reference symbol. Reference lines of 1 m are provided on the X, Y, and Z axes, with the origin at the face of the transducer.

The Dalles Dam, spring and summer 1996

The application of the tracking transducer at The Dalles Dam was intended for determining fish trajectory as affected by manipulations of flow in front of turbine and sluiceway intakes. The Dalles Dam turbine gallery is divided into two sections, the greatest being the main units and the other being two downstream fish units. The study focused on two locations: location 1 between Fish Unit (FU) 2–2 and Main Unit (MU) 1–1, and location 2 between and Main Unit 2–3 and Main Unit 3–1. The tracking transducer was positioned on the piernose face at a depth of 9.5 m at location 1 and at a depth of 10.5 m at location 2. Location 1 was sampled during the spring study, and location 2 was sampled during both the spring and summer studies. Figure 4 shows a three-dimensional view of the transducer at location 1 in relation to the fish unit and main unit structures. The transducer at location 2 was similarly situated.

The Dalles Dam experiment's intention was to create a zone of transition by covering the top half of the turbine entrance so that fish might avoid entering turbines. Data from the tracking transducer were categorized into periods of turbine occlusion (upper half occluded versus completely open turbine entrances), by time period (spring or summer and day or night), by location (location 1 between FU 2–2 and MU 1 or location 2 between MU 2–3 and MU 3–1), and by three different target strength intervals (−55 to −45 dB, −45 to −35 dB and above −35 dB). Location 1 was further blocked into data in front of the fish unit and data in front of the main unit. Data were blocked into day and night periods. The beginning of day and night was taken from a computer program for sunrise and sunset (from *Astronomy*, April 1984:75–77).

The range of compensated target strength in which the system would select a new target to track was −55 dB to −30 dB. Once selected, the target's signal strength on subsequent pings was ignored. As at the Ice Harbor Dam experiment, the target range criterion originally was a mini-

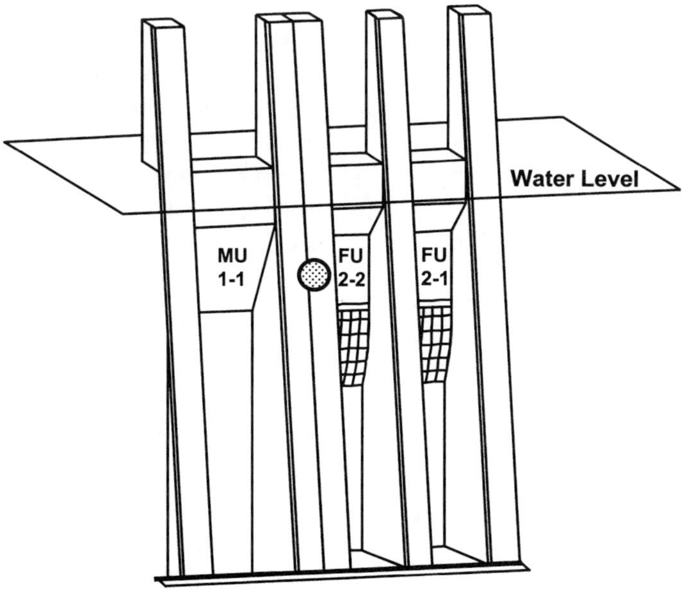

Figure 4. Three-dimensional view of the transducer mounting location at The Dalles Dam location 1 in relation to the fish unit (FU) and main unit (MU) structures. The transducer at location 2 between main units 2 and 3 was similarly situated.

mum distance (radius) from the transducer, between 1.0 and 15.0 m; targets outside this range would not be selected for tracking. The position criteria also included the same Cartesian (X, Y, Z) limits as at Ice Harbor, however positive values were right (approximately to the east), away (upstream, approximately north), and up, respectively. In practice, the volume sampled was limited to a 12 m radius around the transducer, at which distance the −17.5 dB (one way) sidelobes of the transducer encountered the surface.

Similar tracking parameters used at Ice Harbor Dam were used here to define tracking errors if:

- the target density was too high, the system was unable to tell which target matched the one tracked on the previous ping;
- the second closest target was less than three times as far away as the first choice, the track was ended;
- the speed of the tracked target was greater than 4.0 m/s, the track was ended; and
- a target could be tracked immediately following the end of the previous track, a mark was added to the file to show that the best possible target continued to be tracked.

During operation, the transducer angles were systematically set to a new randomized pointing direction every 10 min. The split-beam echosounder was again operated effectively at 5 pings/s.

John Day Dam, spring and summer 1997

Overflow weirs were placed into Spillways 18 and 19 as an experiment to attract fish to surface flows at the spillways and to attempt to prevent them from entering turbines. The fish behavior study determined direction, velocity, path, and vertical and horizontal placement of fish within the water column in front of these spillways. Two BioSonics 420 kHz DT6000 split-beam systems were used for the tracking systems to determine travel routes and velocities of fish within roughly 15 m of the spillway weirs. The tracking systems were lowered about 18 m below the surface, resting on the spillway ogee below anticipated fish passage routes. The BioSonics circular split-beam transducers transmit a 6° beam (full beamwidth at half power) with −30 dB sidelobes (one way). Three clear advances were made over the two previous studies in preparing the tracking transducer system for the John Day study. First, the serial link between the rotator control computer and echosounder control computer was used to control the pinging of the echosounder. This increased ping rate capability and thus tracking capability by a factor of two. Second, predictive fish following was added to the rotator control program. The algorithm predicted incremental movement in Δx, Δy, and Δz separately using the following equation (shown for Δx):

$$\Delta x = \frac{(x_i - x_{i-1}) + 0.5(x_{i-1} - x_{i-2}) + 0.25(x_{i-2} - x_{i-3}) + 0.125(x_{i-3} - x_{i-4})}{1.875} \quad (5)$$

Third, the 420 kHz signals are quieter than those at 120 kHz, and the beam had much lower sidelobes allowing better performance near structures.

The tracking transducer's primary duty is to follow individual fish. The resulting data are measured or apparent fish positions from which velocity vectors, $\bar{V}_{apparent}$, can be estimated as the change in position divided by the change in time. $\bar{V}_{apparent}$ is the sum of the water velocity, $\bar{V}_{water}$, and the "real" fish vector (i.e., fish effort vector), $\bar{V}_{effort}$. The relationship is defined as

$$\bar{V}_{apparent} = \bar{V}_{effort} + \bar{V}_{water},$$

from which the fish effort vector can be estimated as

$$\bar{V}_{effort} = \bar{V}_{apparent} - \bar{V}_{water}.$$

The x, y, and z components of the fish effort vector follow:

$$V_{x,effort} = V_{x,apparent} - V_{x,water},$$

$$V_{y,effort} = V_{y,apparent} - V_{y,water}, \text{ and}$$

$$V_{z,effort} = V_{z,apparent} - V_{z,water}.$$

Water velocity was measured at various stations in a physical model of the John Day Dam located at the Waterways Experiment Station, Vicksburg, Mississippi. It has three full bays (and an additional one half on each side). Velocities were estimated with and without bulkheads (weirs) in bays 18 and 19 at a pool elevation of 80.5 m, and stop openings (left to right) 2, 6, 6, 2, 2. However, the stop opening should have been 3, 4.5, 4.5, 6, and 7 looking downstream left to right. Therefore, the velocity measurements at Spillbays 18 and 19 were corrected by multiplying by 4.5/6. Because the model data were taken at Spillbay 18, their mirror image was used for Spillbay 19 velocity estimates. Sample measurements were then extrapolated to estimate the water velocity vector at a particular fish position in three-dimensional space.

Results

In addition to tabulated and descriptive results, fish tracks at all locations were graphically examined in three dimensions using AutoCAD, and example figures are presented for the John Day Dam.

Ice Harbor Dam

The system was monitored for several weeks, from 26 May until 20 June. The numbers of fish tracked with 5 pings or more were 424 when the sluiceway was off and 466 fish when the sluiceway was on.

Table 1 presents averages of fish tracking data organized by sluiceway condition and in three depth ranges: 0–5 m, 5–10 m and greater than 10 m. Statistics averaged were 1) speed, 2) time tracked, 3) distance tracked, 4) velocity factor, 5) depth, 6) target strength, and 7) unit direction vectors.

Average fish speeds were the same in the two sluiceway conditions, 0.6 m/s. However in the sluiceway on condition, speed was highest in the upper water column (0–5 m) and higher than the sluiceway off condition (0.8 versus 0.7 m/s). The velocity factor showed that the fish meandered more at the surface (0–5 m depths), and this behavior was apparent in the AutoCAD representations.

Table 1. Average statistics of fish tracked by a motorized two-axis tracking split-beam echosounder, at Ice Harbor Dam, Washington, during May and June 1995. Statistics are organized by three sluiceway conditions and three depth ranges: 0–5 m, 5–10 m, and greater than 10 m.

	Sluice off				Sluice on (May–June)[a]				Sluice on (June)[a]			
	0–5 m	5–10 m	10 m	*All*[b]	0–5 m	5–10 m	10 m	*All*	0–5 m	5–10 m	10 m	*All*
Speed (m/s)	0.77	0.59	0.62	*0.65*	0.69	0.56	0.73	*0.63*	0.83	0.67	0.57	*0.62*
Time tracked (s)	1.20	1.33	1.80	*1.40*	1.27	1.55	1.50	*1.44*	1.56	1.16	2.00	*1.68*
Distance tracked (m)	0.87	0.75	0.99	*0.84*	0.87	0.83	0.97	*0.86*	1.27	0.75	1.02	*0.94*
Velocity factor[c]	2.11	1.93	1.61	*1.91*	2.22	1.92	1.50	*2.00*	1.28	1.52	1.57	*1.54*
Depth of tracked fish (m)	3.52	7.11	12.65	*7.34*	3.63	6.96	13.31	*6.21*	4.54	8.54	12.00	*10.45*
Target strength (dB)	−47.64	−50.57	−52.86	*−50.25*	−48.57	−49.30	−54.00	*−49.43*	−44.40	−48.13	−49.12	*−48.56*
Unit direction vectors[d]												
X unit	−0.26	0.23	0.02	*0.06*	−0.18	−0.01	0.12	*−0.01*	−0.23	−0.32	0.37	*0.17*
Y unit	−0.89	−0.55	−0.53	*−0.63*	−0.98	−0.61	−0.67	*−0.75*	−0.96	−0.89	−0.78	*−0.87*
Z unit	−0.38	−0.80	−0.85	*−0.78*	0.11	−0.79	−0.73	*−0.66*	−0.14	−0.31	−0.51	*−0.47*
N (number of fish detected)	121	207	96	*424*	139	185	31	*355*	5	39	67	*111*

[a]The May–June period was 26 May through 16 June; the second period was 17 June through 20 June.
[b]"All" represents the average across all depths except for *N*.
[c]Velocity factor is the amount of meandering between the begin and end points of a tracked fish.
[d]Unit direction vectors are represented by Xunit, Yunit and Zunit, where Xunit is positive toward the south, Yunit is positive away from the dam, and Zunit is positive up.

The most striking feature of the average statistics is that the *Z* component of the unit vector in the upper water column was −0.4 for the sluiceway off condition and +0.1 for the sluiceway on condition. This represents a significant downward movement of fish in 0–5 m depths when the sluiceway was closed. In addition there was higher toward-sluiceway movement (*Y* component of the unit vector) when the sluiceway was open (−1.0 versus −0.9) in the upper water column. It is interesting to note that at the end of the study, after 16 June, there was slight downward movement of fish in the 0–5 m depths (Table 1) when the sluiceway was on.

Fish below 5 m depth showed a downward movement regardless of sluiceway condition. The average downward movement was only slightly less when the sluiceway was on in both the 5–10 m depths (−0.79 versus −0.90) and 10 + m depths (−0.66 versus −0.78).

The Dalles Dam

Vector statistics were generated from 3 m (depth) × 3 m (normal to dam face) × data block (width) volumes. These data corresponded to the data blocks: by occlusion plate installation, by diel period, by target strength, by location, and by study period (spring versus summer). A minimum of three fish tracks was required before an average vector was estimated for any block. Vector plots and vector statistics were made to correspond to volume blocked.

In general, when occlusion plates were removed, fish tracks and vectors were more horizontal and toward the dam in the vicinity of turbine intakes. When the plates were installed, fish traveled in a more vertical and downward direction into the turbine intakes. There was little evidence that near-to-dam fish were able to avoid downward trajectory when occlusion plates were installed. In mid-May, fish entered the sluiceway with greater velocity during the daytime than at night. Ancillary data (BioSonics 1997b) showed that daytime sluiceway entrainment was significantly higher than at night.

A strong downriver component was found at all depths. This downstream component was strongest at location 2, between main units 2 and 3. Movement was strongest at the deeper depths. The 4.5 m and 7.5 m deep blocks (3 m to 9 m depths) appear to have the smallest magnitude downstream component. As fish approach the dam they appear to make more effort to avoid the dam. Fish in the 3 m to 12 m depths also appeared to have smaller approach velocities. These observations suggest a zone of transition between fish in the upper 3 m and those in deeper depths. One hypothesis is that fish in the upper 3 m are easily able to enter the sluiceway, while a transition zone exists from roughly 3–9 m from which fish may be drawn below or be able to escape into the upper level.

Larger fish appear to be more capable of moving against the flow, even in close proximity to the turbine intake, as shown for those fish with average TS greater than −35 dB. Smaller acoustically sized fish entered the sluiceways, fish units, and main units. Some fraction of these smaller acoustically sized fish probably avoided entrainment entirely, passing downstream toward spillways.

There was more downstream movement in the upper 6 m at the second location upstream of the fish units. Although there was considerable temporal separation in the data, one interpretation of this finding is that fish passing by the main units are more likely to enter the fish units' sluiceway or fish units.

John Day Dam

The movement or trajectory of fish upstream of Spillbays 18 and 19 was analyzed by plots of apparent fish movement and partitioned by day and night as well as by weir in (Figure 5a) and weir out (Figure 5b). Plots of fish effort vectors were also examined for weir in and weir out conditions. Summary statistics were made by spatially blocking the data into 3 m × 3 m cross-sectional volumes. The study was divided into three periods: spring (5 May–6 June), summer (7 June–11 July) and extension (12 July–24 July), 1997.

The tracking transducer recorded apparent fish movement that represents the sum of water-induced motion (water velocity) and the fish's swimming movement (fish effort). Thus, apparent movement is governed in part by fish behavior. Tables 2 and 3 present summary statistics for fish tracks for combined day and night periods for the two locations and for the weir in-out conditions.

Fish velocities toward the spillway were highest in the spring and considerably lower in summer and extension periods. Fish were distributed higher in the water column during the spring. They tended to be higher in the water column in the weir in condition than in the weir out condition during the spring and summer. Fish movement had greater toward surface components during the spring and summer. During the spring there was a larger toward Washington-side movement in both weir in and out conditions, but only Spillbay 19 data showed this tendency during the summer period.

Table 3 shows that the number of fish tracked (normalized by the hours of system operation) was influenced by the weir placement. During the spring, the number tracked per hour was highest with the weir in place. However, the number tracked per hour was smaller with the weir in during the summer. During the spring and summer, 70–80% of the fish were moving toward the spillway regardless of weir placement. In both periods the percentage in the upper water column was higher with the weir in, but the highest percentage occurred in spring. The longest track was 27 m, during the extension period.

Fish effort vectors (Figure 6b) were estimated using average fish apparent movement vectors (Figure 6a) and subtracting water velocity from apparent velocity. Figure 6 shows the fish vectors in 3 m × 3 m blocks in transverse section at Spillbays 18 and 19 for the spring period. Fish moved downward and away from the weir in the upper water, with some indication of moving with the flow close to the weir. Fish appeared to try to move away from the spillway within 9 m of the weir slot (weir removed)

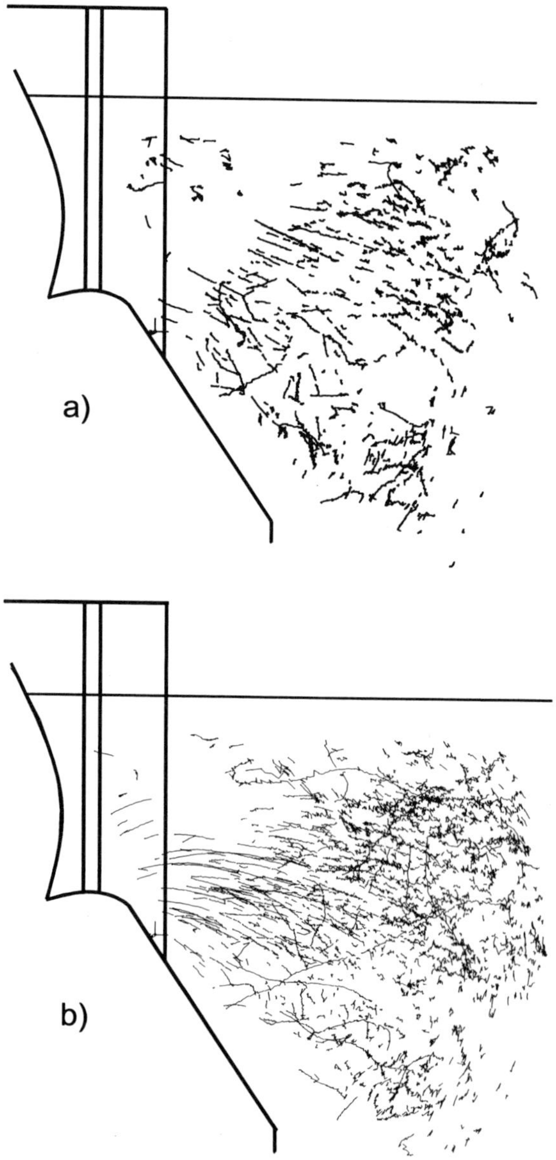

Figure 5. Side view of fish travel paths at John Day Dam Spillbays 18 and 19, in summer 1997. The fish paths are shown for (a) overflow weir installed and (b) overflow weir removed.

Table 2. Summary of tracked fish velocities and positions for spring (5 May to 6 June), summer (7 June to 11 July) and extension (12 July to 24 July), 1997, at John Day spillbays 18 and 19 for weir in and weir out conditions. Data are the mean velocities and positions of apparent fish vectors. Velocity (*Vx*, *Vy*, *Vz*) is positive in *x* toward the Oregon side, in *y* upstream from the spillbay, in *z* toward the surface. *X*, *y*, *z* are measured from the transducer, which was located 18 m below the surface.

	N	*Vx* (m/s)	*Vy* (m/s)	*Vz* (m/s)	*x* (m)	*y* (m)	*z* (m)
			Spring				
18, In	571	−0.13	−0.42	0.05	−2.04	9.87	8.06
18, Out	152	−0.17	−0.41	0.02	−2.57	9.57	6.45
19, In	293	−0.10	−0.31	0.03	2.10	9.73	7.79
19, Out	195	−0.23	−0.37	0.00	0.96	9.30	4.63
			Summer				
18, In	515	0.01	−0.19	0.08	−2.91	17.06	4.93
18, Out	2,094	0.05	−0.22	0.07	−2.49	17.19	2.70
19, In	1,016	−0.01	−0.17	0.07	2.14	15.64	2.41
19, Out	2,252	−0.09	−0.23	0.05	1.63	16.30	1.14
			Extension				
18, In	973	−0.04	0.03	0.09	−2.75	17.72	3.95
18, Out	722	0.01	−0.09	0.08	−3.05	18.07	4.99
19, In	1,256	−0.11	0.00	0.06	1.67	18.33	1.64
19, Out	819	−0.05	−0.06	0.08	2.19	18.28	2.49

and, farther away, to actively swim toward the dam. In both the summer and extension periods, fish also moved downward and away from the dam near the surface although with more meandering farther away from the dam and less active swimming toward the dam.

Table 3. Summary of tracked fish directions overall and in the upper 8 m for spring (5 May to 6 June), summer (7 June to 11 July) and extension (12 July to 24 July), 1997, at John Day spillbays 18 and 19 for weir in and weir out conditions. To weir, upper 8 m, and upper 8 m to weir are proportions of total fish found 1) moving toward the spillbay, 2) in the upper 8 m water column, and 3) in the upper 8 m moving toward the spillbay. Sigma (dB) is the mean fish target strength averaged in intensity space.

	N	Effort (h)	*N*/effort	To weir	Upper 8m	Upper 8m to weir	Sigma (dB)
			Spring				
18, In	571	93.2	6.1	0.88	0.49	0.42	−46.29
18, Out	152	45.2	3.4	0.88	0.42	0.36	−46.74
19, In	293	85.8	3.4	0.79	0.48	0.36	−46.17
19, Out	195	88.1	2.2	0.86	0.23	0.15	−48.55
			Summer				
18, In	515	59.0	8.7	0.75	0.20	0.15	−46.75
18, Out	2,094	128.1	16.4	0.84	0.16	0.12	−48.48
19, In	1,016	135.7	7.5	0.72	0.21	0.15	−47.28
19, Out	2,252	206.9	10.9	0.75	0.12	0.07	−48.29
			Extension				
18, In	973	45.7	21.3	0.53	0.22	0.13	−45.23
18, Out	722	47.0	15.4	0.62	0.24	0.13	−46.11
19, In	1,256	61.8	20.3	0.54	0.16	0.08	−45.96
19, Out	819	52.0	15.8	0.62	0.17	0.09	−45.71

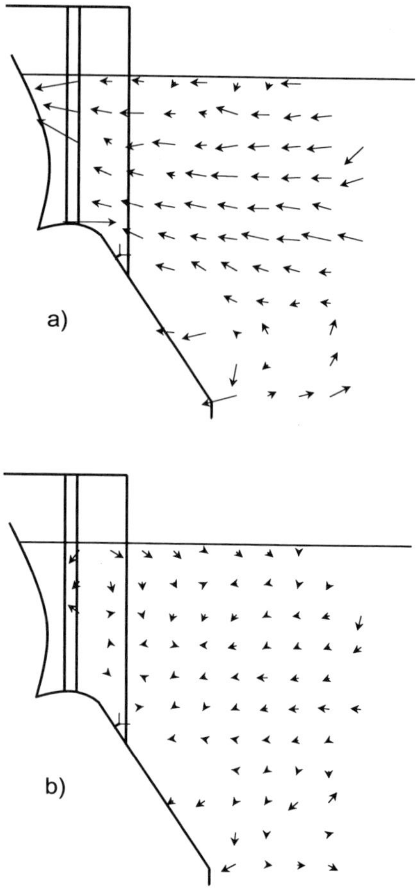

Figure 6. Side view of average fish velocities in 3 m x 3 m blocks at Spillbays 18 and 19 for the 1997 spring period at John Day Dam spillway. The fish effort velocity vector can be estimated as . Apparent fish velocities are shown in (a) and estimated fish effort velocities are shown in (b), when an overflow weir was installed.

Fish were distributed at all depths within the set range limit of the tracking transducer systems, roughly 25 m upstream of the piernose face. The effect of the flow downstream toward spillbays on apparent fish movement appeared fairly consistent with distance upstream in the spring and summer periods, when movement downstream is prevalent. In the extension period however, fish appeared to be milling. The fish effort vectors showed that the fish moved more actively away from the dam in the extension period than in spring and summer, thus holding position. The extension period may include larger nonsalmonid species capable of moving against flow in front of the spillbays. The fish track plots from the three periods indicated fish vertical distribution is generally unaffected by day and night conditions, although fish movement appears more motivated during the day and more aligned with streamlines at night.

Discussion

The tracking transducer methodology offers a way to "see" fish movement over longer periods of time than is possible with conventional methods. Some track lengths at the John Day Dam were nearly 30 m. These kinds of close range data would be impossible to acquire with conventional fixed systems.

The tracking limitations of the tracking sonar are based in stepper motor angular velocity, the ping rate of the split-beam sounder, fish speed, and the detection limits of the split-beam transducer. The Simrad transducer was limited by a maximum one-way gain compensation of 6 dB and thus about $\pm 4.95°$ of beamwidth. The BioSonics transducer had slightly larger aperture ($\pm 5.7°$) and was constrained by phase "wraparound." Pinging at 5 pings/s, the maximum fish movement and thus angular tracking rate is 25°/s, assuming the target is initially on-axis. The recent advance to 10 pings/s increases angular tracking speed by two (about 50°/s).

The stepper motor control software used a weighted prediction. It received the alongships angle and athwartships angle measurements, then programmed the stepper motors to keep the main axis of the transducer beam aimed on the target, thereby tracking the target. Future application can use other predictive trackers such as Kalman filtering (Beard et al. 1994). Predictive tracking increased the capability to track fish behavior, both lengthening recorded tracks and allowing faster speeds to be recorded.

A tracking transducer system can provide many benefits over the conventional fixed-beam transducer system. The ability to track fish as they move through the water while centering the fish on the acoustic axis can provide specific information regarding the fish behavior and target strength. This ability to continuously track fish can be applied toward many research and management applications. Some examples include separating large prey from baitfish, predator–prey interaction studies, direction of movement near and around structures (oil platforms), riverine enumeration studies, and behavioral studies around structures such as hydroelectric dams and natural and artificial reefs.

At Ice Harbor Dam's 4B sluiceway and 1.8 m wide vertical slot configuration, the tracking transducer showed that fish meandered at the surface but did so less and approached the slot when the sluiceway was open. This was true in the upper 5 m of water. Below 5 m there was significant movement into the turbine intakes.

The tracking transducer shows advantages over fixed systems (for example Ransom and Ouellette 1988) for behavior observation. They suggested that fish were found in depths of 1–2 m while we show fish are distributed down to and past the level of the turbine intake at Ice Harbor Dam. The tracking transducer data compare well with the present mobile survey data that found an abundance of fish in the 4–5 m depths (BioSonics 1997b). This new tool suggests that the fish tend to move into the sluiceway as long as they remain above 5 m. Hydrodynamic theory suggests that streamlines present in the upper 5 m near the dam may be shallower away from the dam. Thus fish found at 4–5 m could be entrained if they are passive and do not move to the upper streamlines.

At the Dalles Dam, the tracking transducer performed best in deeper depths. As a result, there were less data for fish traveling at depths shallower than the bottom of the sluiceway intake. This was especially true during sampling at the second location between main units 2 and 3. The majority of fish tracks were taken just below the level of the sluiceway intake at location 2. Near-surface noise was one reason for the smaller number of fish tracks close to the surface. Transducers with low sidelobes should be used in future studies in similar acoustic noise backgrounds, for example at John Day Dam. Recently (August 1998) the predictive tracking algorithm has been extended several pings into the future. Adult salmon were tracked with tracks nearly 50 m long from a fixed barge on the Fraser River.

Currently, the system has the capability of tracking a single fish over a time period. In the future, it is hoped that more than one fish can be tracked by utilizing predictive tracking and taking advantage of the extremely high speed of the stepper motor system. The tracking transducer can move in excess of 140°/s. One half of a hemispherical volume can be scanned in less than 10 s with a ping rate of 10 pings/s.

Much of the 1995 behavioral study effort at Ice Harbor Dam was spent on developing the tracking split-beam system. It became immediately apparent that the approach offered considerable advantages. In particular, the tracking split-beam system offers the potential to bridge the gap between radio tracking and conventional fixed location acoustics. Radio tracking provides long tracks of fish movement. However, because of the effort and cost involved the sample size is small. Conventional fixed location acoustics provide detailed information on relative rates of passage at specific points, with very large sample sizes. However, it is apparent from the observed milling behavior, the horizontal differences in both abundance and diel behavioral patterns, and other intriguing diel patterns of fish passage, that the approaches and near-field behavior can be complex and that behavior may provide critically important information with regard to the effectiveness of surface biological collectors.

Nontracking split-beam systems can provide directional information but only over the same limited dimensions as conventional fixed-location acoustics. In contrast, tracking split-beam offers the possibility of providing intermediate track lengths and, especially, detailed behavioral information in the near field that is not only valuable in its own right but can considerably enhance the value of the radio tracking and conventional fixed-location information. We also successfully tracked 200 kHz acoustic tags with the tracking transducer (Johnson et al. 1998), and this new capability should be valuable in augmenting the array of tools to study and evaluate fish bypass facilities.

Acknowledgments

The authors would like to thank the U.S. Army Corps of Engineers' Walla Walla and Portland Districts for supporting innovative approaches for studying fish behavior responses to dam structures.

References

Beard, G. C., T. G. McCarter, W. Spodeck, and J. E. Fletcher. 1994. Real-time acquisition and tracking system with multiple Kalman filters. Pages 202–213 *in* M. K. Masten, L. A. Stockum, M. M. Birnbaum, and G. E. Sevaston, editors. Acquisition, tracking, and pointing VIII. Proceedings of SPIE volume 2221. SPIE (International Society for Optical Engineering), Bellingham, Washington.

BioSonics. 1996. Acoustic evaluation of the surface bypass and collection system at Ice Harbor Dam in 1995. Final Contract Report to the Walla Walla District, U.S. Army Corps of Engineers, Seattle.

BioSonics. 1997a. Hydroacoustic evaluation and studies at the Dalles Dam spring/summer 1996. Volume 2-smolt behavior. Final Report prepared for the U.S. Department of the Army, Portland District, Seattle.

BioSonics. 1997b. Hydroacoustic evaluation and studies at the Dalles Dam spring/summer 1996. Volume 1. Final Report prepared for the U.S. Department of the Army, Portland District, Seattle.

Jaffe, J. S. 1995. FTV, a sonar for tracking macrozooplankton in three-dimensions. Deep-Sea Research 42:1495–1512.

Johnson, R. L., and six coauthors. 1998. Behavioral acoustic tracking system (BATS). Report prepared for U.S. Army Corps of Engineers, Walla Walla District, Walla Walla, Washington.

MacLennan, D. N., and E. J. Simmonds. 1992. Fisheries acoustics. Fish and fisheries series 5. Chapman and Hall, London.

Sherman, S. M. 1984. Monopulse principles and techniques. Artech House, Norwood, Massachusetts.

Ransom, B., and D. Ouellette. 1988. Hydroacoustic evaluation of juvenile fish passage at Ice Harbor Dam in spring 1987. Contract Report, BioSonics, Inc. Seattle.

Thorne, R. E., and G. E. Johnson. 1993. A review of hydroacoustic studies for the estimation of salmon downriver migration past hydroelectric facilities on the Columbia and Snake rivers in the 1980s. Reviews in Fisheries Science 1(1):27–56.

Martin Mallen-Cooper

Developing Fishways for Nonsalmonid Fishes: A Case Study from the Murray River in Australia

12

In the last one hundred years there have been dramatic declines in the range and abundance of native freshwater fish in southeastern Australia. A major contributing factor to these declines has been the inhibition or prevention of fish passage at more than 1,500 dams and weirs. Additionally, the few fishways that were built at these barriers were based on salmonid designs and were thus not suitable for the Australian fish fauna.

To redress the situation, new fishways were designed and built based on laboratory tests with native species using an experimental vertical-slot fishway. The present study concerns the assessment of one of these fishways at Torrumbarry Weir on the Murray River. The wide size range (120–600 mm) of fish ascending the fishway and the reduction in fish density below the weir after the fishway was operational, indicated the success of the fishway. The dominant portion of golden perch *Macquaria ambigua* (Percichthyidae) and silver perch *Bidyanus bidyanus* (Terapontidae) migrating upstream were immature fish. Field experiments defined diel movement patterns, ascent time, and movement through an experimental tunnel, and demonstrated that the swimming ability of one species in situ was greater than estimates from laboratory experiments.

This study emphasizes the significance of four important steps in developing fishways: 1) identifying the species and life stages (and sizes) that are migrating, 2) testing these fish in an experimental fishway, 3) designing and building the fishway, and 4) quantitatively assessing the fishway. Fishways for nonsalmonid fishes have frequently failed because steps 1, 2, and 4 have been ignored.

Introduction

Of the many impacts of river regulation on fish populations, one of the most striking is the disruption of migratory pathways. For many salmonid populations barriers to upstream migration of adult fish have been mitigated with considerable success using fishways (Clay 1995). However, apart from some designs for anadromous clupeids in North America (e.g., Richkus 1974) and France (Larinier 1990), fishways for nonsalmonid fishes have frequently failed (Petts 1984; Welcomme 1989).

In the Murray-Darling river system of southeastern Australia, the impact of dams and weirs on fish migrations is often cited as one of the major causes of the dramatic decline in the range and abundance of freshwater fish (Lake 1971; Cadwallader 1978; Brumley 1987). The migratory nature of native fish in this river system has been known and utilized for harvest by indigenous people for tens of thousands of years (Dargin 1976). All of the 33 species in this drainage need free passage along streams to some extent and 14 species are known to make large-scale movements (Mallen-Cooper 1989).

The need for fishways was recognized early in the development of river regulation in Australia (Stead 1914). In the Murray-Darling river system, 22 fishways were built from 1930 to 1989 (Mallen-Cooper 1989), although this is a small number in proportion to the over 1,500 dams and weirs estimated to be on the main streams of this system (New South Wales Department of Land and Water Conservation, unpublished data). Unfortunately, the fishways were based on salmonid designs from the Northern Hemisphere and were not suitable for the native fish fauna, which is entirely nonsalmonid (Mallen-Cooper and Harris 1990). The slopes of these fishways, commonly between 1:8 and 1:5, were also steeper and thus more turbulent than those frequently used for salmon (Mallen-Cooper 1989).

Initial research on fishways in Australia focused on the passage of fish within coastal fishways in eastern Australia (Kowarsky and Ross 1981; Beumer and Harrington 1982; Russell 1991). One report for the Murray-Darling river system summarized fish passage through one fishway (Langtry 1940, in Cadwallader 1977). Harris (1984) reported that the fishways in coastal southeastern Australia were poorly maintained and suggested that the designs were ineffective for native fish. These studies did not assess the number of fish attempting to migrate nor did they manipulate water velocities within these fishways to assess the swimming ability of the fish.

Following advice from Eicher (1982), the suitability of a vertical-slot fishway was investigated using laboratory models (Mallen-Cooper 1992, 1994). At least two native species, golden perch *Macquaria ambigua* (Percichthyidae) and silver perch *Bidyanus bidyanus* (Terapontidae) from the Murray-Darling river system could ascend this type of fishway if water velocities were less than typical velocities for salmonid designs.

Based on design criteria developed in the laboratory, eight new vertical-slot fishways were built. The first in the Murray-Darling river system was completed at Torrumbarry Weir on the Murray River in February 1991. Previously, the weir had prevented fish from getting access to over 350 km of upstream river habitat. The fishway is the longest—at 131 m—in Australia; it contains 38 pools and surmounts a 6.5 m high weir.

The objectives of the present study were to assess the effectiveness of this new vertical-slot fishway, and to validate and refine the laboratory data in the field. In so doing, it provides a case study for developing fishways for nonsalmonid fishes.

Methods

Study area

The Murray-Darling river system (Figure 1) has a large catchment of 1.073×10^6 km^2, with a relatively small mean annual discharge of 10,090 gigalitres (GL). Annual flow is highly variable; from 1894 to 1993 it ranged from 1,626 GL to 54,168 GL (Maheshwari et al. 1995). Most of this discharge originates from the headwaters of the Murray River and 10% comes from the semiarid Darling River. The Murray River is regulated by dams in the upper catchment, 14 weirs along its middle and lower reaches, and tidal barrages at the mouth of the river (Jacobs 1990).

Torrumbarry Weir is situated in the middle reaches of the Murray River well above its confluence with the Darling River (Figure 1). During large floods the weir is dismantled and removed from the river. The nearest weirs on the main channel are 354 km upstream at Yarrawonga, and 528 km downstream at Euston (Eastburn 1990). The reach downstream of Torrumbarry is the longest free-flowing reach in the river. Torrumbarry Weir is also downstream of major irrigation offtakes (Jacobs 1990), and as

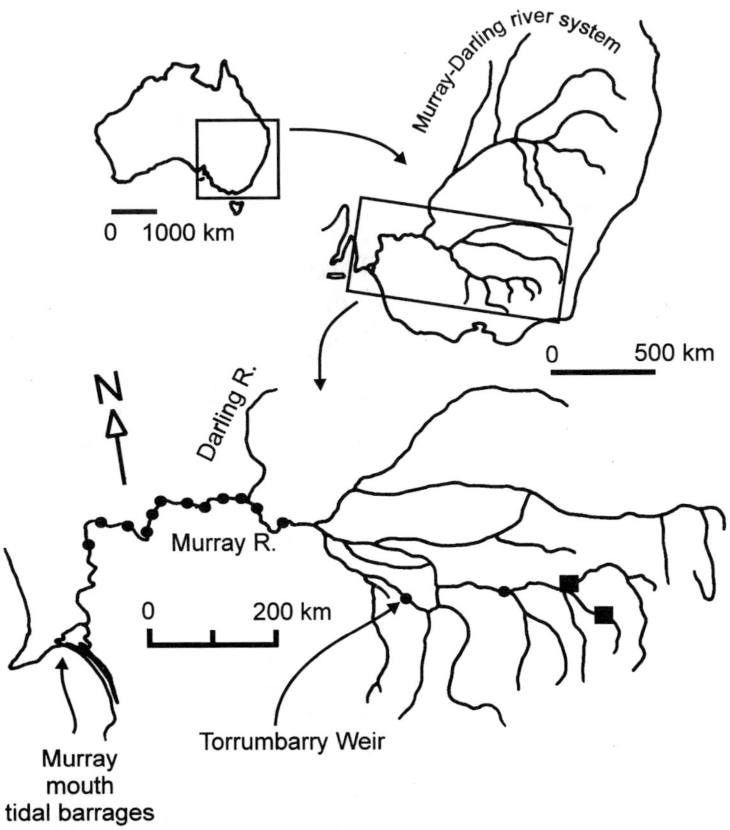

Figure 1. The Murray-Darling river system of southeastern Australia and the location of Torrumbarry Weir and fishway. Weirs and dams on the Murray River are shown by solid circles and squares, respectively.

such the streamflow downstream of the weir does not suffer the inverted seasonality of more regulated upstream sites. Instead, the river retains some of its natural flow regime of high flows in spring and low flows in autumn, although total monthly flows are much reduced (Close 1990).

Description of fishway

Torrumbarry fishway is a pool-type fishway with vertical-slot baffles. The fishway is 2 m wide and 1.1 m deep for most of its length, and its course resembles an inverted "Z." The pools are 3 m long by 2 m wide and the baffles are similar to the "Seton Creek" single-slot design in Clay (1995), with 0.3 m wide slots. Further details are provided by Mallen-Cooper (1993). The head difference between pools is 0.165 m providing a maximum water velocity of 1.8 m/s, which was derived directly from laboratory experiments (Mallen-Cooper 1994). Turbulence, or energy dissipation, in the pools averages 105 W/m^3, calculated as in White and Penino (1980) in units of power.

Torrumbarry Weir has a concrete apron on the downstream side, which is a barrier to fish movement at low tailwater levels. To alleviate this problem, the fishway has two entrances: one at the downstream edge of this apron to accommodate low tailwater levels and one adjacent to the weir to accommodate high tailwater levels when the apron is submerged. The lower pools of the fishway are also deeper (up to 1.83 m) to accommodate some tailwater variation. The lower third of the fishway, on the concrete apron, is connected to the higher entrance with a 22 m long, 1.5 m wide channel with no baffles.

Fishway performance: distribution and passage of migratory fish

River sampling

To assess the effect of the fishway on the distribution of migratory fish, three sites were selected in the Murray River: 35 km above Torrumbarry Weir, immediately (<1 km) below Torrumbarry Weir, and 6 km below Torrumbarry Weir. The most upstream site was chosen to be above the weir pool and in flowing water habitat comparable to the two downstream sites. The site immediately below the weir was chosen to detect fish migrating upstream and accumulating below the weir. The most downstream site was chosen to avoid any local effect of the weir on fish moving upstream and so provide a comparison with the other two sites.

The three sites were sampled monthly between January 1990 and June 1992, on separate and contiguous days. Six gill nets (two replicates of 38, 62, and 100 mm stretched mesh) and two fyke nets (two replicates of 20 mm square mesh) were used at each site. Each replicate of nets was set on opposite sides of the river and within 200 m; the nets were evenly spaced along each bank and the sequence of nets was random. Nets were set for 14–17 h, from 2 h before dusk to at least 2 h after dawn. All captured fish were identified and measured for fork length, or total length for the two round-tailed fishes (*Macquaria ambigua*, Murray cod *Macullochella peelii peelii*). Catch rates were standardized to 14 h by the proportional change in sampling time. Temperature, dissolved oxygen, and river flow were recorded at each site. The sampling regime provided standardized catch rates between the three sites and an estimate of relative fish density. The completion of the fishway in February 1991 provided 13 monthly netting samples before, and 17 samples after, the fishway became operational.

There were three hypotheses associated with this sampling:

H_1: Before installing the fishway and during migratory periods, there will be higher relative densities of fish immediately below the weir compared to 6 km downstream.

H_2: Before installing the fishway, there will be lower relative densities of fish 35 km above the weir compared with the site 6 km downstream of the weir.

H_3: After the fishway is operating there will be no difference in the relative fish density between the three sites.

To identify significant differences in relative fish density between the three netting sites, a two-way analysis of variance (ANOVA) of log-transformed data using time (month) and site (35 km above, <1 km below, and 6 km below weir) as factors was used. To identify which site has the greatest relative fish abundance ($P < 0.05$) for each species in each month, one-way ANOVAs were used followed by least significant difference (LSD) tests.

Monthly two-day sample of the fishway entrance

The fishway entrance was sampled to help identify fish that were migrating upstream during the sampling with nets, provide a check on the selectivity of the nets, and identify fish that were not able to ascend the fishway. A single cone-trap was used at the entrance for two periods of 22 h, concurrent with the monthly netting samples. Water flow through the fishway was reduced so that the head loss at the entrance baffle was lowered from 165 mm to 80–100 mm. This enabled smaller fish to enter, potentially providing a more comprehensive sample of the migratory fish community.

Daily sampling of fish reaching the top of the fishway

A cage, 2 m square and 1.1 m deep with a cone-trap, was installed at the top of the fishway. Beginning in February 1991, permanent staff (Goulburn-Murray Water) at the weir monitored fish collected in the trap each weekday; the present study includes data up until June 1992.

I also collected independent samples of fish entering the fishway and ascending the fishway on paired days when high numbers of fish were migrating. One-tailed Kolmogorov–Smirnov tests for large samples were used to compare the size distribution of each species at the two locations in the fishway.

Fishway performance: experimental manipulations of the fishway

General method

Four aspects of fish behavior (escapement from cone-traps, passage though a tunnel, diel movement, and fallback) and three aspects of swimming ability (maximum water velocity negotiated, velocity negotiated in a channel, and ascent time) were investigated by experimental manipulations of the fishway to assess its performance and validate the earlier laboratory data. The experiments were done between January 1993 and March 1993.

The general method was to dewater the fishway and remove all fish, set up an experimental condition (e.g., tunnel, diel period) in the fishway and then run the experiment allowing fish to enter freely from the river. Cone-traps prevented fish from leaving the fishway once they had entered and the fish were not handled in any way until the end of each experiment, when all fish were identified, counted, and measured. The fishway was dewatered to collect fish and all cone-trap exits were covered with wire mesh before dewatering to prevent increased escapement from stressed fish. To reduce variances, statistical analysis within an experiment was limited to species with at least five individuals in each replicate.

This method was developed as preliminary experiments in the fishway showed that to obtain realistic responses from fish they could not be captured and transported to the fishway. For most species, screens also could not be used to accumulate fish below the fishway.

Escapement from cone-traps

Escapement of fish was tested for two types and locations of cone-traps:

a) The cone-trap within the cage at the top of the fishway. The opening of the cone was 300 mm wide × 1,400 mm high and tapered over 1.3 m to a 100 mm wide by 350 mm high exit. The cage and cone-trap were covered with 25 mm square mesh but were later additionally covered with 20 × 15 mm "birdwire." The cage was set in the still water of the impoundment.

b) The cone-trap within the fishway. This had a 530 mm wide × 900 mm high opening tapering over one meter to a 180 mm wide × 330 mm high exit. This was covered in 20 × 15 mm "birdwire." A single cone-trap and a double cone-trap using two pools of the fishway were tested.

Fish were allowed to enter the fishway for 24 h and pass through the cone-trap being tested. Flat screens were then placed across the vertical slot, two pools below the cone-trap. Fish were cleared from these two pools using a two-pole dip-net that was the width of the channel. Once cleared, the test period began. At the end of each period, fish were cleared from the same two pools, representing escaped fish, and from above the cone-trap, representing fish that did not escape.

The cone-trap in the cage was tested for escapement of fish over 20 min, which was applicable to the ascent time trial; and over 22 h, which was applicable to the standard daily monitoring of the fishway. The single and double cone-traps in the fishway were tested for periods of 5 h, corresponding to the shortest time period in the swimming ability experiments; and 23 h, longer than the maximum periods in the swimming ability and tunnel experiments.

Passage through a tunnel

To test the effect of tunnels on the passage of fish, three pools of the fishway were covered with black plastic creating a 9-m-long tunnel with no detectable light in the middle pool. Single cone-traps were placed one baffle below the tunnel, to capture fish that approached the tunnel but did not enter, and at the upper end of the tunnel. Fish entering the fishway were exposed to the tunnel for 18.5 h, enabling fish species moving in different diel periods to be sampled. The lower cone-trap was then covered with wire mesh and the fishway was run for an additional 1.5 h, enabling fish that had just entered the fishway to attempt to negotiate the tunnel. Otherwise, these fish would be included in the sample of fish that were unable to negotiate the tunnel.

As a control, the same procedure was done with no tunnel, that is, with cone-traps but no black plastic. These two treatments, tunnel and no tunnel, were tested on consecutive days and randomized within the two-day block. The block was replicated three times and a paired *t*-test was applied to the log-transformed data.

Diel movement and fallback

The fishway entrance was monitored with a single cone-trap for three days during five different periods of the day: dawn (2 h), morning (till midday), afternoon (after midday), dusk (2 h), and night. Light intensity (Lux) was measured with a light meter at the beginning and end of each diel period. Catch data were standardized for each species and for each diel period, to the daily proportion of fish per hour. That is, the number of fish within a diel period was divided by the total number of fish for the day and by the number of hours in that period.

I also observed some bony herring *Nematalosa erebi* moving back down the fishway at the end of daylight. To quantify this movement I trapped bony herring entering the fishway on three separate days; then at night I checked the number that had ascended the fishway and the number that had moved back down the fishway.

Maximum water velocity negotiated

The maximum water velocity negotiated by fish ascending the Torrumbarry fishway was tested by changing the head loss (or step height) of water across one baffle within the fishway. Head loss is a major criterion used in designing fishways, and water velocity is directly related to head loss by the following equation (Vennard and Street 1982):

$$V = Cd.(2g.\ \Delta h)^{0.5}$$

where V = velocity (m/s)
Cd = coefficient of discharge (1.0 [Clay 1995])
g = acceleration due to gravity (9.8 m/s^2)
Δh = head loss (m)

The head loss across a baffle was changed by adjusting the flow of water using a dewatering gate at the top of the fishway and by changing the slot width (to a minimum of 130 mm) with plywood and a clamp. Head loss was measured to ±2 mm with piezometer tubes.

Two cone-traps were placed above and below the test baffle and the water velocity adjusted. This procedure could take 30 min, so the exit of the lowest cone-trap was covered with wire mesh to prevent fish entering the fishway while water velocities were low. Once the correct water velocity was achieved, the wire mesh was removed from the lowest cone-trap and fish could enter the fishway for 16–19 h. (In the laboratory experiments only 20 min was used but this proved insufficient in the field.) At the end of this period the wire mesh was replaced on the lowest cone-trap, and the fishway continued to run for 2 h, similar to the tunnel experiment.

To accommodate the diel movement period of bony herring, the experiments were done during daylight, and the initial sampling period was reduced to 3 h. This was followed by the same 2 h period with the lowest cone-trap covered.

Probit analysis was used to model the proportion of fish negotiating each water velocity tested. Low-velocity data points in which less than 100% of fish negotiated the fishway baffle were excluded from the analysis, if 100% of fish had negotiated a higher velocity. The assumption here is that if fish can negotiate a higher velocity then it is some factor other than swimming ability that results in fewer fish negotiating a lower velocity. To assess if reducing the slot width reduced fish movement, the minimum slot width (130 mm) was tested at a low water velocity, where swimming ability was not limiting fish passage.

Water velocity negotiated in a channel

To assess the water velocities negotiated by fish within the 22 m long connecting channel, a Marsh McBirney 201D current meter was used at one cross-section in the middle of the channel. Water velocity was measured at three depths (surface, middle, bottom) at each of three points (left hand edge, middle, right hand edge).

Ascent time

During a bony herring migration, I monitored ascent time during the day. After clearing the fishway of fish, the two downstream slots at the fishway entrance were blocked with flat screens for 1 h to attract bony herring to the entrance. The screens were then lifted and the cage at the top of the fishway was checked every 30 min for 6 h. Removal of fish from the cage took 2.25 min, and a flat screen covered the end of the fishway during this period to prevent ascending fish escaping into the impoundment. The 6-h procedure was repeated three times over three days.

Results

Species composition and distribution in the river and fishway

A total of seven native species and four nonnative species were collected (Table 1). All species caught in the river used the fishway except for one nonnative species, goldfish. Redfin perch did not use the fishway during standard sampling, but high numbers of juvenile redfin perch entered the base of the fishway during experiments in 1993.

Fish abundance above and below the weir and fish passage through the fishway

Golden perch, silver perch, bony herring, carp, and redfin perch showed very significant differences in relative densities with time ($P < 0.0001$) and, excepting redfin perch, between sites within the same month ($P < 0.0001$). Relative fish density was lower in the upstream site (35 km above) compared to the downstream site (6 km below) on only one occasion (January 1990) for one species (golden perch). The only other significant differ-

Table 1. Total number of fish collected in the nets (three sites combined) from Jan 1990 to Jun 1992 and passing through the fishway (in 5 d per week) from Feb 1991 to Jun 1992.

Species	Number of fish	
	Nets	Fishway
Native		
Golden perch *Macquaria ambigua* [Percichthyidae]	765	2,095
Silver perch *Bidyanus bidyanus* [Terapontidae]	394	3,391
Bony herring *Nematalosa erebi* [Clupeidae]	95	432
Murray cod *Maccullochella peelii peelii* [Percichthyidae]	13	31
Shortheaded lamprey [a] *Mordacia mordax* [Mordaciidae]	1	3
Catfish [b] *Tandanus tandanus* [Plotosidae]	0	1
Australian smelt *Retropinna semoni* [Retropinnidae]		abundant in base of fishway
Totals	1,268	5,953
Nonnative		
Common carp *Cyprinus carpio* [Cyprinidae]	739	3,219
Redfin perch *Perca fluviatilis* [Percidae]	393	0
Goldfish *Carassius auratus* [Cyprinidae]	162	0
Brown trout *Salmo trutta* [Salmonidae]	1	1
Totals	1,295	3,220

[a] Also known as Australian lamprey.

[b] Also known as dewfish.

ences in catches of native fish involved higher catches less than 1 km below the weir, compared to 35 km above and 6 km below the weir. Figure 2 displays catches in the downstream sites and the fishway for the three most common species, combined with environmental data.

The fishway was completed and began operating in February 1991. The base of the fishway could be used for sampling from November 1990. However, the cone-trap was moved by large carp, once in November 1990 and again in December 1990, which left only one sample each for these two months. Subsequently, the cone-trap was held in place with clamps.

High relative densities of silver perch, golden perch, and carp immediately below the weir were recorded in February, October, and November of 1990, prior to completion of the fishway (Figure 2). There were only minor increases in the density of silver perch and golden perch below the weir after the fishway was operating. From December 1990 to April 1991 the low densities of these two species and carp below the weir coincided with high numbers of these species using the fishway. Carp in December 1991 provide the only sample where there was a high density of fish below the weir with fish using the fishway.

For silver perch and carp in November 1990 and for carp in October 1991 there were high numbers below the weir with very low numbers entering the fishway. During these two monthly samples there were also high flows which submerged the lower half the fishway.

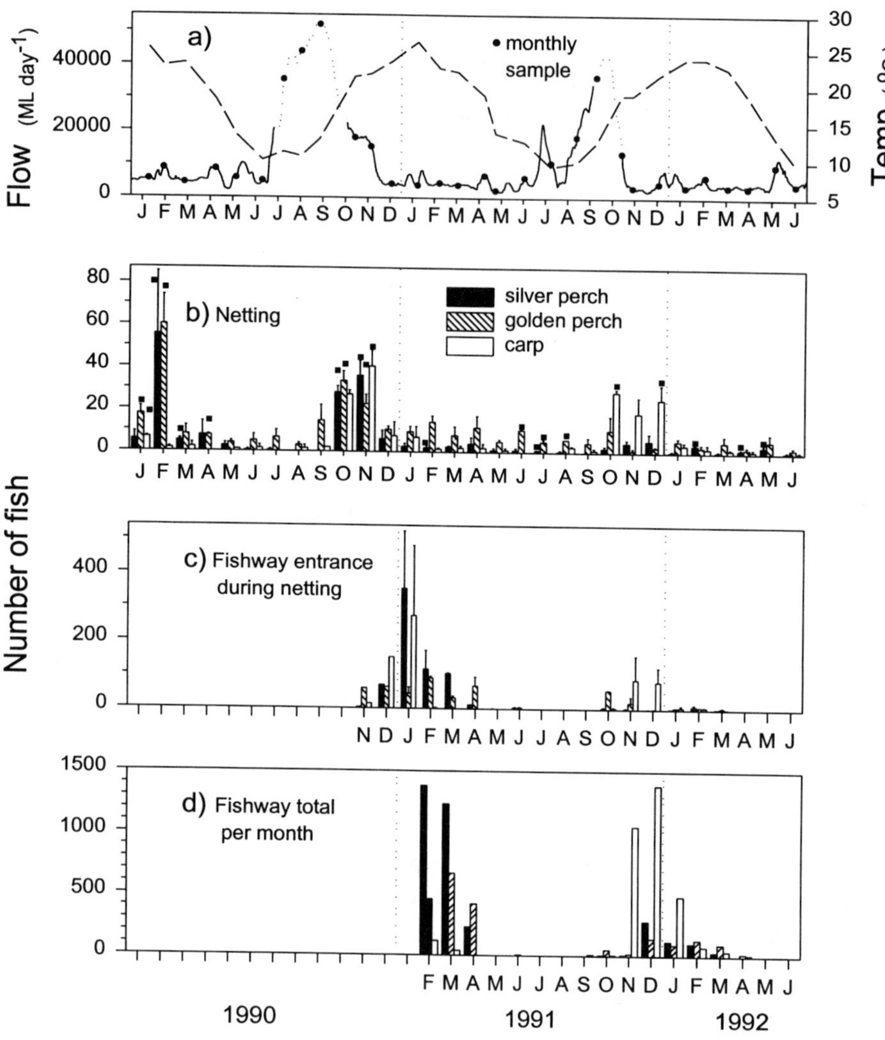

Figure 2. Sampling at Torrumbarry fishway. Graph a) depicts river flow (dotted line represents the period the weir was removed from the river), water temperature, and sampling times (solid circles). The other graphs indicate the numbers of silver perch, golden perch, and carp collected: b) in nets immediately below the weir (mean ±SE, $N = 2$) (species in each month that had a significantly greater catch rate than the site 6 km downstream are marked with a square symbol); c) entering the fishway on two days during netting samples (mean ±SE, $N = 2$; only one sample in November and December 1990); and d) passing through the fishway during weekdays each month.

Significantly higher densities of redfin perch were found immediately below the weir, compared to the other two sites, in 5 of the 30 monthly samples. This species did not, however, enter the fishway on these occasions. Goldfish were occasionally collected in nets below the weir, but no fish used the fishway. Bony herring were collected in variable low numbers below the weir, but consistent numbers used the fishway from October 1991 to March 1992.

Fish migration and river flow

Golden perch, silver perch, and carp migrated upstream during relatively low flows and moderately high flows. For example, these species entered the fishway in high numbers in January 1991 (Figure 2c) when the flow was 3,796 ML/d (exceeded 67% of the time on a flow duration curve for the period of record, data from the Murray-Darling Basin Commission) and were accumulating below the weir in October 1990 when the flow was 18,025 ML/d (exceeded 25% of the time).

Comparison of species and size of fish at the top and bottom of the fishway

Figure 3 shows the size distributions of fish collected in the top and bottom of the fishway. Silver perch between 80 and 400 mm fork length (FL) entered the fishway at the bottom, but fish between 80 and 120 mm FL were not recorded at the top. Silver perch less than 130 mm in length were able to pass through the mesh of the cage at the top of the fishway.

For golden perch and bony herring, all size classes that entered the fishway were well represented at the top of the fishway. There were, however, significantly ($P < 0.01$) fewer smaller fish of each species at the top of the fishway. The majority of silver perch and golden perch using the fishway were immature fish, whereas most of the bony herring were mature fish.

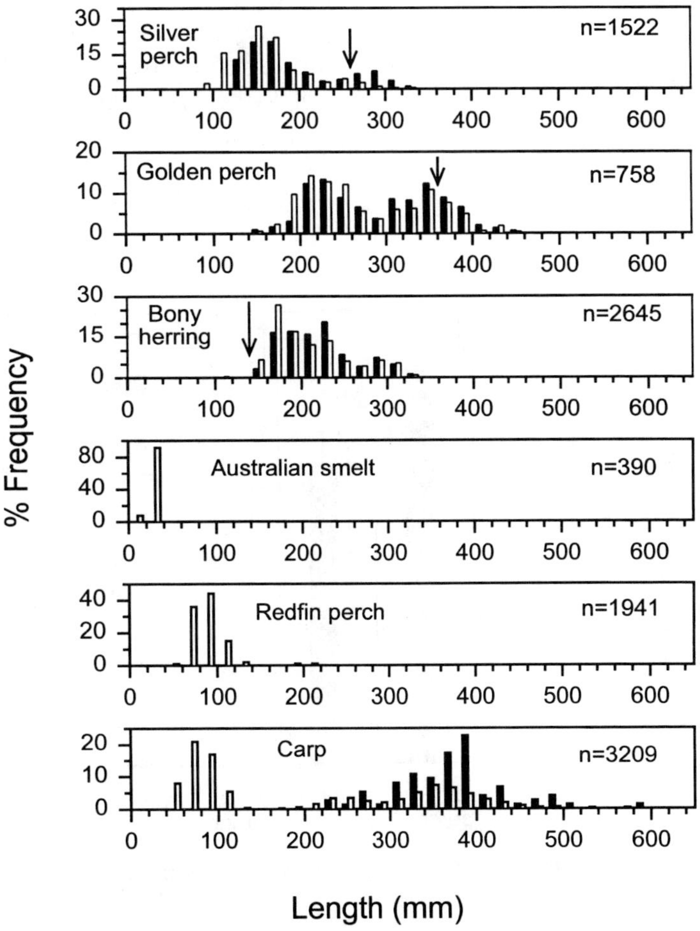

Figure 3. Size distribution of fish collected at the top (solid bar) and bottom (open bar) of the fishway. Minimum size at maturity is indicated by an arrow for golden perch, silver perch (Mallen-Cooper 1996), and bony herring (Puckridge and Walker 1990).

Australian smelt were observed actively entering the fishway in high numbers, but they could only ascend the first few pools of the fishway. A sample taken with a fine-meshed dip-net showed these fish were mostly less than 40 mm FL (Figure 3). Redfin perch were mainly less than 140 mm FL and were unable to ascend the fishway. Two major size classes of carp attempted to ascend the fishway; the large fish (220–520 mm FL) reached the top, but the small fish (40–120 mm FL) did not.

Fish behavior in the fishway

During escapement trials the most abundant species using the fishway was bony herring. Mean escapement rates of three replicate experiments for this species were between 0 and 2.5% ($n = 32 - 116$) for all combinations of traps and time periods, except for the single cone-trap set for 24 h where the rate rose to 5%. A maximum of four fish escaped from the single cone, one fish from the double cones over 23 h, and one fish from the cage. Results for golden perch and silver perch were unreplicated, but no escapement was recorded from the cage or cone-traps.

In the tunnel experiment, three native species, golden perch, silver perch, and bony herring, and two nonnative species, redfin perch and goldfish, used the fishway. The passage of golden perch through the fishway was unaffected by the tunnel; the passage of silver perch was reduced by a mean of 13% (SE 2%) ($P < 0.01$); and bony herring were completely excluded (Figure 4). For the nonnative species, redfin perch was almost completely excluded from passing through the fishway with a tunnel, but carp were unaffected. For these two species, most of the fish were small (60–100 mm), and although high numbers entered the fishway (between 706 and 1,631/d), few ascended successfully when no tunnel was present.

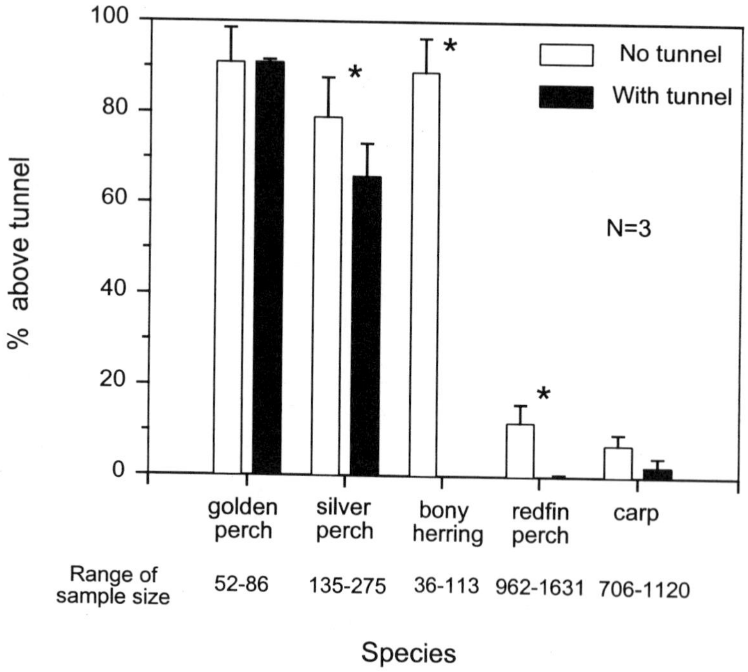

Figure 4. Proportion of fish passing through the same 9 m section of fishway with and without a tunnel (created by black plastic). Significant differences are indicated by an asterisk.

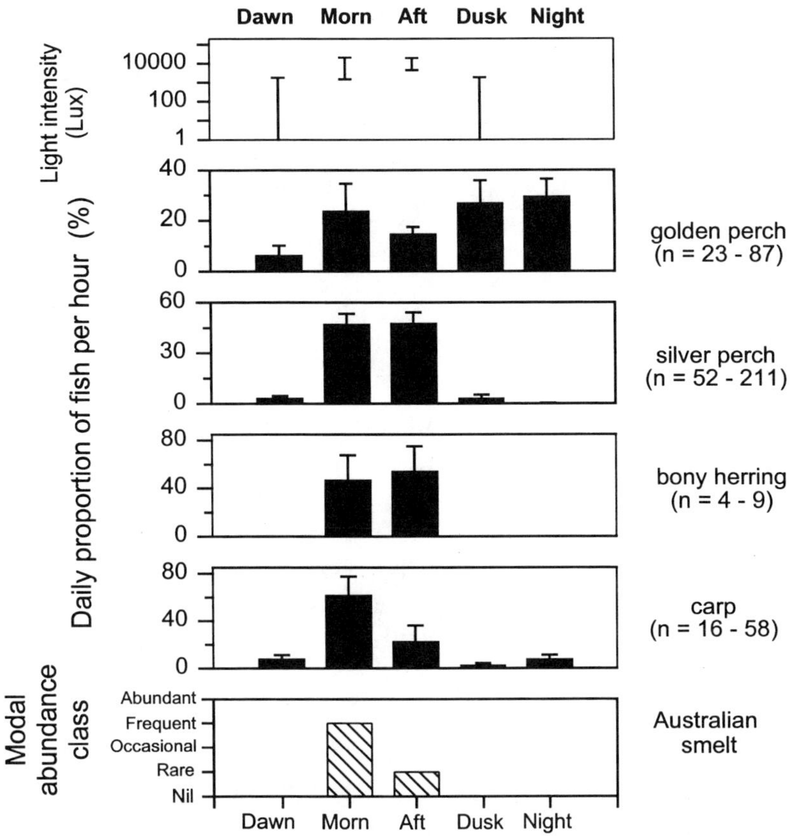

Figure 5. Diel upstream movement of five species entering the fishway on three consecutive days (mean ± SE). The range of total numbers per day for each species is shown in brackets.

In the diel movement study, adult bony herring, immature smelt, and immature silver perch moved upstream only during daylight hours (Figure 5). Carp moved upstream mainly during daylight, and immature golden perch moved upstream during all diel periods. In the fallback study between 36 and 44% of the bony herring that entered the fishway moved back down the fishway at the end of daylight, if they had not completed their ascent. Other species were not migrating during this assessment, so their behavior is unknown.

Swimming ability in the fishway

During the study of the maximum water velocity that fish could negotiate, golden perch, silver perch, and bony herring were migrating upstream. For each species the maximum water velocity that 95% of fish negotiated is shown in Figure 6 and summarized in Table 2. For all three species there were some low-velocity trials where less than 100% of fish negotiated the fishway baffle even though 100% had negotiated a higher velocity. For golden perch and silver perch there was no inhibition to pass through a narrow (130 mm) slot. However, an escapement of only one fish in each trial (indicated by a vertical error bar in Figure 6) would reduce the low velocity data points from 100%. There was no recorded escapement for bony herring, but some fish did not pass through the narrow slot at a low test velocity (Figure 6). Such behavior would contribute to variance in the data.

Table 2. Summary of the maximum water velocities negotiated by 95% of fish (NV95) across a single vertical-slot baffle in the fishway; laboratory results for golden perch are included (Mallen-Cooper 1994).

	Fish length (mm) (mean ±SD)	NV95 (m/s)	95% confidence limits	
			Lower	Upper
Golden perch	276 ± 95	2.53	2.35	2.63
Golden perch (lab.)	441 ± 16	1.83	1.43	2.03
Silver perch	229 ± 102	2.70	2.54	2.77
Bony herring	228 ± 45	2.64	2.21	2.76

During normal operation of the fishway, all fish collected at the top during low tailwater levels must pass through the 22 m long connecting channel. Water velocities in the channel ranged from 0.20 m/s on the edges to 0.33 m/s in the middle of the channel. Any fish that could ascend the lower fishway appeared able to negotiate the channel.

During the three trials on ascent time, bony herring (228 mm mean FL) started arriving at the top of the fishway after 1.5–3 h had elapsed (Figure 7). On each of the three days there was a consistent rate of fish ascending the fishway. The minimum ascent time was between 1.5 and 2.5 h, corresponding to a rate of 2.3–3.8 min per pool.

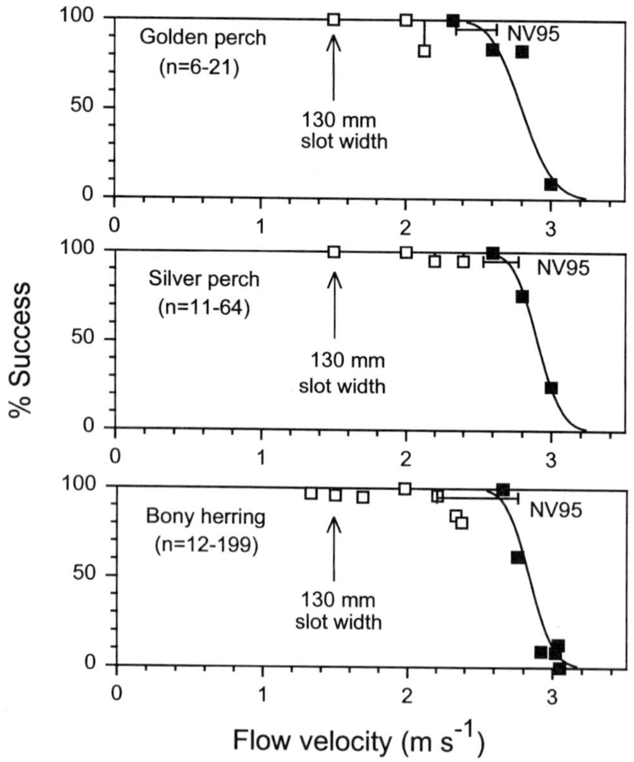

Figure 6. The proportion of fish negotiating different water velocities through one vertical-slot baffle in the Torrumbarry fishway. Probit regressions and the NV95 with 95% confidence limits are shown for each species. Open data points were not used in the probit analysis, and the vertical lines from these points indicate potential escapement of one fish from the cone-traps.

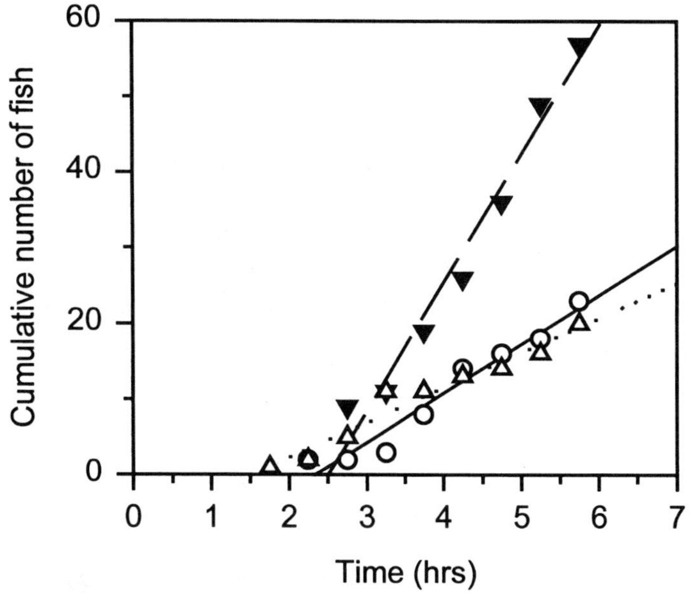

Figure 7. The time of arrival of bony herring at the top of Torrumbarry fishway, after all fish have been cleared from the fishway. Data for three separate days are shown.

Discussion

Fishway assessment

One measure of the effectiveness of a fishway is the relative density of fish in the river immediately downstream, compared to the number of fish passing through the fishway. Fish accumulated below the Torrumbarry Weir before the fishway was completed, but once the fishway was operating there were only minor aggregations of fish below the weir at low to moderate flows while large numbers of fish passed through the fishway. This indicated that fish could find the fishway entrance, pass through a vertical-slot baffle, negotiate the water velocities and turbulence within the fishway, and the fishway had the capacity to pass the numbers of migrating fish.

One species of fish, Australian smelt, and one observed invertebrate, a Paratya shrimp, were unable to ascend the fishway. They were, however, both commonly observed above the weir before construction of the fishway. The role of upstream migration in their life history is unclear.

Torrumbarry fishway appeared to perform poorly at higher flows when it was partly submerged. Fish were migrating at these flows and other studies have shown golden perch in the Darling River migrating at very high flows (Mallen-Cooper 1996). The migration of fish at low flows at Torrumbarry indicates that fishways need to be designed for a wide range of flows in the Murray-Darling River system, at least until evidence suggests this is not necessary to sustain these migratory populations.

In assessing the Torrumbarry fishway, the upstream riverine site provided little additional information to the downstream sites. Possibly more extensive sampling would have revealed an increase in the numbers of upstream migrants above the weir after the fishway was installed. However, in a short-term assessment of a single fishway where upstream migration is the focus, I suggest that using only localized sites downstream is likely to be more efficient. Beyond immediate changes in fish distribution, a greater test of a fishway's effectiveness lies in the long-term sustainability of the migratory fish populations along the whole river, and a broader sampling strategy would be necessary.

Sampling the top and bottom of the fishway proved to be a useful technique in assessing the fishway's efficiency. It identified that the majority of size classes of native fish that entered the fishway could ascend the full length, except for Australian smelt. The absence or restriction of some small size classes of golden perch, bony herring, and possibly silver perch, although statistically significant, may not be ecologically significant when the bulk of migrating immature and mature fish can ascend the fishway. It could even be argued that this represents an effective compromise in the design of the fishway. To pass the smaller fish the fishway would need to have lower water velocities and turbulence. This would result in a longer, less steep, and more expensive fishway.

The compromise between fish passage and capital cost is faced for most fishways for both nonsalmonid and salmonid fishes. The present data could form the basis of a valuable performance criterion in the design and assessment of fishways. For example, "the fishway needs to be able to pass at least 95% of the size range of each migratory life stage of each species."

Redfin perch and small carp were unable to ascend the fishway. Unfortunately, larger carp extensively used the fishway, and designing a fishway that limits the spread of this nonnative species while allowing native migratory fish to pass, does not seem readily possible. On the plains of the Murray-Darling river system, most weirs are either submerged at some time during floods or have gates lifted to pass floodwaters, and thus already permit the spread of nonnative fish.

The number of fish passing through the Torrumbarry fishway might be considered high, but there is potential for much greater numbers. Up to 3,000 golden perch per day attempt unsuccessfully to ascend the next fishway 512 km downstream, which is an old salmonid design (Mallen-Cooper 1996). An effective fishway at this downstream weir would very likely increase the numbers of migratory fish at Torrumbarry.

Records of fish passage from the top of a fishway are widely used to indicate a fishway's effectiveness (e.g., Jowett 1987 in New Zealand; Quirós 1989 in South America; Pavlov 1989 in the former USSR; Travade 1990 in France; Lonnebjerg 1990 in Denmark; Bok 1990 in South Africa; Zhili et al. 1990 in China; Katopodis et al. 1991 in Canada; Russell 1991 in Australia; Cada and Sale 1993 in the United States; and Sato et al. 1995 in Japan), but there is a fundamental problem in defining an ecologically significant number of fish. In the present study, estimates of fish passage would not have identified fish accumulating below the weir or small fish species or size classes of fish unable to ascend the fishway. The numbers of migrating fish at Torrumbarry are also greatly influenced by the effectiveness of the downstream fishway. Therefore, to accurately assess the performance of a fishway it would seem essential to have quantitative measures of the migratory fish community as it approaches, enters, and ascends the fishway.

Assessment is an essential component of developing fishways for migratory species where there is little knowledge of the behavior of these fishes in fishways. Assessment provides essential feedback to refine designs. In Australia the lack of quantitative assessment allowed the construction of unsuitable fishways to continue for 70 years (Mallen-Cooper 1996).

Fishway experiments

In attempting to validate the earlier laboratory experiments on swimming ability, the same techniques were unsuccessful in the field; it appeared that if fish were handled in any way, they stopped migrating upstream and some moved back downstream. Similar results have also been reported for other fishway experiments where fish are handled and transported (Fisk 1959; Anonymous 1967; Nakamura et al. 1995). Another reason for the poor results of the laboratory techniques in the field may be that the fish in the laboratory were ripe adult fish, and the fish in the field were almost all immature subadult fish. The latter may not have the same inclination to move upstream.

The in situ techniques developed for the field are similar to those used by Monk et al. (1989) and avoided handling fish until after the experiment. The disadvantages are an unknown sample size or species composition in each trial and less information about the period of time fish negotiated a particular treatment in the fishway. But the great advantage is that fish continue to migrate, or attempt to migrate, up the fishway. These field techniques may apply to other species and other river systems, particularly where migration is opportunistic or facultative in contrast to obligate, such as those of semelparous anadromous salmon.

An experimental fishway or baffle is particularly useful to initially determine whether the design suits the behavior of the fish. The use of a single baffle and pool to test swimming ability in situ is also an economical method. To test swimming ability through many vertical-slot baffles, nine to eighteen pools are needed with an adjustable slope to maintain consistent head losses between pools (Rajaratnam et al. 1986). Such long, full-scale models of fishways are possible in a large in situ laboratory like the Fisheries-Engineering Research Laboratory on the Columbia River (e.g., Collins and Elling 1960; Monk et al. 1989). The method used in the present study can be adapted to short, full-scale in situ models of two or three pools, or it can be used at the downstream end of existing fishways by modifying baffles. Both of these methods are relatively inexpensive.

The NV95 (the maximum velocity that 95% of fish can negotiate) for golden perch was much greater in the field compared with the laboratory where fish were captured and transported. The fish in the field were much smaller in body size, however, and would be expected to have a poorer swimming ability.

The greater NV95 could be the result of the extra time given for fish to negotiate the test velocity (18–21 h compared with 20 min in the laboratory experiments). However, considering bony herring needed only 2.3–3.8 min to pass through a single pool and baffle, and that larger golden perch in the laboratory needed 20 min to do the same, it seems more likely that the laboratory experiments underestimated the swimming ability of golden perch.

If the original laboratory results had been accurate, a higher water velocity would likely have been recommended and the resulting fishway would have excluded many smaller fish. It was certainly fortunate that the laboratory data were conservative and hence the fishway allowed the passage of the smaller fish. For golden perch and silver perch, these smaller fish are immature, and they compose the dominant portion of the population that is migrating upstream. This is in contrast to the supposition that mainly adults of golden perch and silver perch migrated upstream

(Lake 1971; Llewellyn and MacDonald 1980; Reynolds 1983; Merrick and Schmida 1984; Battaglene 1991), which led to the choice of adult fish for the earlier laboratory experiments. This highlights the importance of identifying the migratory fish community first.

In the field experiments on swimming ability, silver perch and bony herring were of similar size as the golden perch, and the NV95s were similar. The high water velocities negotiated by these fish may imply that future fishways could be built with steeper slopes and therefore would be cheaper to build. However, ascent time of fish and diel movement patterns of different species are also critical factors in the design of effective fishways. For example, bony herring moved upstream only during daylight and needed 1.5–2.5 h to reach the top of the fishway. If these fish did not make it to the top by the end of the day they returned back down the fishway, resulting in over one-third of the fish being unsuccessful each day. Even though most bony herring are capable of negotiating 2.6 m/s, they need a significant period of time to negotiate a number of pools and slots with a maximum velocity of 1.8 m/s.

Diel movement patterns in migrating freshwater fish are common. North American clupeids (Talbot 1952; Richkus 1974) and salmonids (Clemens 1958; Brawn 1979; Blackett 1987; Andrew 1990) are known to migrate upstream mainly during daylight, as do some cyprinids and a cichlid in South Africa (Cambray 1990). Catostomids, percids, esocids (Schwalme et al. 1985; Pavlov 1989; Katopodis et al. 1991), and some cyprinids (Silva and Davies 1986) move upstream during specific diel periods in addition to daylight.

Early observations of the diurnal movements of salmon led to concern that these fish might move back down a fishway at night (Andrew and Geen 1960). I am unaware of any published accounts where this prediction was fulfilled for salmonids. It is, however, an accurate prediction for the behavior of bony herring in the Torrumbarry fishway and may also be true for the numerous other migratory nonsalmonid species that move upstream in specific diel periods. In designing fishways for nonsalmonid fishes, it would seem prudent to investigate diel movement patterns.

In summarizing the field experiments, the velocity criteria of a fishway should not be solely a function of the swimming ability of fish; diel movement patterns, ascent time, and the length of the fishway should also be considered. With this knowledge the Torrumbarry fishway has now been redesigned with large resting pools, like many salmonid fishways, to prevent fallback of fish.

The experiments on swimming ability also provide a guide to how future data might be interpreted for fishway design. The NV95, if accurate, would appear too high to use directly for fishway design. However, if the velocity criterion of 1.8 m/s were considered acceptable for the fish negotiating the Torrumbarry fishway, then an NV95 (for one fishway baffle) reduced by approximately 30% might be appropriate for a 38-pool fishway.

The inhibition and inability of some native species to pass through a tunnel has implications for fishways for nonsalmonid fishes. Reluctance of fish to enter or pass through tunnels in fishways has also been reported for Atlantic herring *Clupea harengus* (Clupeidae) in rivers draining into the Black Sea (Pavlov 1989) and for some fish in Pakistan including the clupeid genus *Hilsa* (Ahmad et al. 1962). Interestingly, pool-type fishways for clupeids in North America have been problematic (Collins 1951; Jackson 1952; Talbot 1952; Monk et al. 1989). Monk et al. (1989) reports that American shad *Alosa sapidissima* do not pass through submerged orifices in fishways. It is possible that a reluctance of this species to pass through tunnels might be the cause of this behavior. Successful fishways for this species such as the Denil design, or the modified pool and weir design developed by Monk et al. (1989) have a path through the fishway that is continuously open to the water surface with no tunnels. Avoiding tunnels when developing fishways for nonsalmonid fishes would appear to be an appropriate cautious measure.

This recommendation could also be applied to the design of road culverts that need to pass nonsalmonid fishes. High water velocities can also be a problem for fish passage in road culverts. The water velocities of 0.20–0.33 m/s from the 20 m long connecting channel of the Torrumbarry fishway provide an initial guide for the species and size of fish migrating in the Murray River and possibly for other similar nonsalmonid fishes.

Developing fishways for nonsalmonid fishes

The success of the fishway at Torrumbarry indicates that a vertical-slot design with relatively low head losses between pools and low turbulence has potential for a range of nonsalmonid fishes. The assessment of the fishway and the outcomes of extrapolating the laboratory data present a case study which identifies four steps that should be considered in developing fishways for nonsalmonid fishes:

1. Identify the migratory fish community, and diel movement patterns.
2. Test fish in an experimental fishway. (In situ experiments are recommended, with no handling of fish until the end of each trial.)
3. Design and build the fishway in the following manner:
   - Extrapolate swimming speed data conservatively;
   - If the range of flows over which fish are migrating is unknown, then design the fishway to operate over the widest possible range;
   - Avoid tunnels;
   - Incorporate resting pools in long fishways, especially if fish exhibit strong diel movements; and
   - Maintain flow and temperature regimes that stimulate fish migration. Although not considered in this paper, it would be remiss to omit this aspect which has caused the failure of otherwise well-designed fishways (e.g., Baras et al. 1994)
4. Assess the fishway.
   - Use quantitative and relevant performance criteria.
   - Assess the migratory fish community that approaches, enters, and ascends the fishway.
   - Do not depend solely on counts of fish passage.

These steps are important in the development of successful fishways for any species. The most common strategy with fishways for nonsalmonid fishes in the past has been to build the fishway and avoid steps 1, 2, and 4. With a more comprehensive approach we can be more optimistic about the future of fishways for nonsalmonid fishes.

Acknowledgments

The construction of Torrumbarry fishway was funded by the Murray-Darling Basin Commission (MDBC). This research was funded by the Natural Resources Management Strategy of MDBC (Project No. N002), NSW Fisheries, and the Cooperative Research Center for Freshwater Ecology. I am grateful to Fergus Hides-Pearson and Ivor Stuart for extensive technical assistance in the field, in addition to Stuart Curran, Simon Hartley, Tim Marsden, and Garry Thorncraft. I thank John Harris for supervision of the work and for comments on the manuscript. I thank Terry Holt, Alan Williams, and Peter Klowss for monitoring the fishway and helping with the fieldwork. Dennis Reid and Geoff Gordon, from the Fisheries Research Institute, provided valuable statistical advice. I would like to thank Brian Lawrence, Norm Mackay (MDBC), and Stuart Rowland (NSW Fisheries) who supported the project and gave important advice. Lastly, I thank Jane Mallen-Cooper for valuable editorial comments on the manuscript.

References

Ahmad, M., C. M. Ali, and S. Ahmad. 1962. Designing of fish ladders. West Pakistan Irrigation Research Institute, Lahore, Technical Report 362/HYD/1962, Lahore.

Andrew, F. J. 1990. The use of vertical-slot fishways in British Columbia, Canada. Pages 267–274 *in* S. Komura, editor. Proceedings of the international symposium on fishways '90 in Gifu. Publications Committee of the International Symposium on Fishways '90, Gifu, Japan.

Andrew, F. J., and G. H. Geen. 1960. Sockeye and pink salmon production in relation to proposed dams in the Fraser River system. Bulletin of the International Pacific Salmon Fisheries Commission 11.

Anonymous. 1967. Annual report (1966-67). Delta fish and wildlife protection study report 6. Resources Agency of California.

Baras, E., H. Lambert, and J. Philipart. 1994. A comprehensive assessment of the failure of *Barbus barbus* spawning migrations through a fish pass in the channelized River Meuse (Belgium). Aquatic Living Resources 7:181–189.

Battaglene, S. C. 1991. The golden perch, *Macquaria ambigua* (Pisces: Percichthyidae) of Lake Keepit, New South Wales. Master's thesis. University of New South Wales, Sydney.

Beumer, J. P., and D. J. Harrington. 1982. A preliminary study of movement of fishes through a Victorian (Lerderderg River) fish-ladder. Proceedings of the Royal Society of Victoria 94:121–132.

Blackett R. F. 1987. Development and performance of an Alaska steeppass fishway for sockeye salmon (*Oncorhynchus nerka*). Canadian Journal of Fisheries and Aquatic Sciences 44:66–76.

Bok, A. H. 1990. The current status of fishways in South Africa and lessons to be learnt. Proceedings of a workshop on the rationale and procedures for the evaluation of the necessity for fishways in South African rivers. March 1990:87–99.

Brawn, V. M. 1979. Efficiency of an estuarine fishway for the passage of salmon (*Salmo salar* L.), Sheet Harbour, Nova Scotia. Canada Fisheries and Marine Service Report 1523.

Brumley, A. R. 1987. Past and present distributions of golden perch *Macquaria ambigua* (Pisces: Percichthyidae) in Victoria, with reference to releases of hatchery-produced fry. Proceedings of the Royal Society of Victoria 99:111–116.

Cada, G. F., and M. J. Sale. 1993. Status of fish passage facilities at non-federal hydropower projects. Fisheries 18(7):4–12.

Cadwallader, P. L. 1977. J. O. Langtry's 1949-50 Murray River investigations. Fisheries and Wildlife Paper 13, Victoria, Australia.

Cadwallader, P. L. 1978. Some causes of the decline in range and abundance of native fishes in the Murray-Darling River system. Proceedings of the Royal Society of Victoria 90:211–224.

Cambray, J. A. 1990. Adaptive significance of a longitudinal migration by juvenile freshwater fish in the Gamtoos River system, South Africa. South African Journal of Wildlife Research 20(4):148–156.

Clay, C. H. 1995. Design of fishways and other fish facilities, 2nd edition. Lewis, Boca Raton, Florida.

Clemens, W. A. 1958. The Fraser River salmon in relation to potential power development. Pages 3–10 *in* P. A. Larkin, editor. The investigation of fish-power problems. Institute of Fisheries, University of British Columbia, Vancouver.

Close, A. 1990. The impact of man on the natural flow regime. Pages 61–74 *in* N. Mackay and D. Eastburn, editors. The Murray. Murray-Darling Basin Commission, Canberra, Australia.

Collins, G. B. 1951. A fishway that shad ascend. U.S. Fish and Wildlife Service Special Scientific Report 65.

Collins, G. B., and C. H. Elling. 1960. Fishway research at the fisheries-engineering research laboratory. U.S. Fish and Wildlife Service Circular 98.

Dargin, P. 1976. Aboriginal fisheries of the Darling-Barwon rivers. Development & Advisory Publications of New South Wales, Dubbo.

Eastburn, D. 1990. The River Murray - history at a glance. Murray-Darling Basin Commission. Canberra, Australian Capital Territory, Australia.

Eicher, G. J. 1982. A fish passage facility programme for New South Wales. New South Wales State Fisheries. Fisheries Research Institute, Sydney.

Fisk, L. O. 1959. Experiments with a vertical baffle fishway. California Fish and Game 45:111–122.

Harris, J. H. 1984. A survey of fishways in streams of coastal south-eastern Australia. Australian Zoologist 21:219–233.

Jackson, G. B. 1952. Passage of shad at the Bonneville fishways. U.S. Fish and Wildlife Service Special Scientific Report Fisheries 94.

Jacobs, T. A. 1990. River regulation. Pages 39–60 *in* N. Mackay and D. Eastburn, editors. The Murray. Murray-Darling Basin Commission, Canberra, Australia.

Jowett, I. G. 1987. Fish passage, control devices and spawning channels. Pages 138–155 *in* P. R. Henriques, editor. Aquatic biology and hydroelectric power development in New Zealand. Oxford University Press, Auckland, New Zealand.

Katopodis, C., A. J. Derksen, and B. L. Christensen. 1991. Assessment of two Denil fishways for passage of freshwater species. Pages 306–324 *in* J. Colt and R. J. White, editors. Fisheries bioengineering symposium. American Fisheries Society, Symposium 10, Bethesda, Maryland.

Kowarsky, J., and A. H. Ross. 1981. Fish movement upstream through a central Queensland (Fitzroy River) coastal fishway. Australian Journal of Marine and Freshwater Research 32:93–109.

Lake, J. S. 1971. Freshwater fishes and rivers of Australia. Nelson, Sydney, Australia.

Larinier, M. 1990. Experience in France: fish pass design criteria and downstream migration problems. Pages 65–74 *in* S. Komura, editor. Proceedings of the international symposium on fishways '90 in Gifu. Publications Committee of the International Symposium on Fishways '90, Gifu, Japan.

Llewellyn, L. C., and M. C. MacDonald. 1980. Family Percichthyidae. Pages 142–149 *in* R. M. McDowall, editor. Freshwater fishes of South-eastern Australia. Reed, Sydney, Australia.

Lonnebjerg, N. 1990. Fishways in Denmark. Pages 253–259 *in* S. Komura, editor. Proceedings of the international symposium on fishways '90 in Gifu. Publications Committee of the International Symposium on Fishways '90, Gifu, Japan.

Maheshwari, B. L., K. F. Walker, and T. A. McMahon. 1995. Effects of regulation on the flow regime of the River Murray, Australia. Regulated Rivers 10:15–38.

Mallen-Cooper, M. 1989. Fish passage in the Murray-Darling Basin. Pages 123–136 *in* B. Lawrence, editor. Proceedings of the workshop on native fish management. Murray-Darling Basin Commission, Canberra, Australia.

Mallen-Cooper, M. 1992. The swimming ability of juvenile Australian bass, *Macquaria novemaculeata* (Steindachner) and juvenile barramundi, *Lates calcarifer* (Bloch), in an experimental vertical-slot fishway. Australian Journal of Marine and Freshwater Research 43:823–834.

Mallen-Cooper, M. 1993. Fishways in Australia; past problems, present successes and future opportunities. ANCOLD (Australian National Committee on Large Dams) Bulletin 93:22–33.

Mallen-Cooper, M. 1994. Swimming ability of adult golden perch, *Macquaria ambigua* (Percichthyidae), and adult silver perch, *Bidyanus bidyanus* (Teraponidae), in an experimental vertical-slot fishway. Australian Journal of Marine and Freshwater Research 45:191–198.

Mallen-Cooper, M. 1996. Fishways and freshwater fish migration in south-eastern Australia. Doctoral dissertation. University of Technology, Sydney, New South Wales, Australia.

Mallen-Cooper, M., and J. H. Harris. 1990. Fishways in mainland south-eastern Australia. Pages 221–229 *in* S. Komura, editor. Proceedings of the international symposium on fishways '90 in Gifu. Publications Committee of the International Symposium on Fishways '90, Gifu, Japan.

Merrick, J. R., and G. E. Schmida. 1984. Australian freshwater fishes: biology and management. Griffin Press, Adelaide.

Monk, B., D. Weaver, C. Thompson, and F. Ossiander. 1989. Effects of flow and weir design on the passage behavior of American shad and salmonids in an experimental fish ladder. North American Journal of Fisheries Management 9:60–67.

Nakamura, S., M. Noda, H. Misawa, and Y. Nakahara. 1995. Temporarily installed long, steep Denil fishway on the Ohkohchi-Sabo Dam in the Abe-Gawa River. Pages 155–162 *in* Proceedings of the international symposium on fishways '95 in Gifu. Publications Committee of the International Symposium on Fishways '95, Gifu, Japan.

Pavlov, D. S. 1989. Structures assisting the migrations of non-salmonid fish: USSR. FAO Fisheries Technical Paper 308.

Petts, G. E. 1984. Impounded rivers - perspectives for ecological management. Wiley, London.

Puckridge J. T., and K. F. Walker. 1990. Reproductive biology and larval development of a gizzard shad, *Nematalosa erebi* (Günther) (Dorosomatinae: Teleostei), in the River Murray, South Australia. Australian Journal of Marine and Freshwater Research 41:695–712.

Quirós, R. 1989. Structures assisting the migrations of non-salmonid fish: Latin America. COPESCAL (Commission for Inland Fisheries of Latin America) Technical Paper 5. FAO, Rome.

Rajaratnam, N., G. Van der Vinne, and C. Katopodis. 1986. Hydraulics of vertical-slot fishways. Journal of Hydraulic Engineering 112:909–927.

Reynolds, L. F. 1983. Migration patterns of five fish species in the Murray-Darling River system. Australian Journal of Marine and Freshwater Research 34:857–871.

Richkus, W. A. 1974. Factors influencing the seasonal and daily patterns of alewife (*Alosa pseudoharengus*) migration in a Rhode Island River. Journal of the Fisheries Research Board of Canada 31:1485–1497.

Russell, D. J. 1991. Fish movements through a fishway on a tidal barrage in subtropical Queensland. Proceedings of the Royal Society of Queensland 101:109–118.

Sato, R., M. Sugawara, K. Otsuka, and H. Aoyama. 1995. Study on upstream migration of fish through a vertical slot fishway. Pages 317–321 *in* S. Komura, editor. Proceedings of the international symposium on fishways '95 in Gifu. Publications Committee of the International Symposium on Fishways '95, Gifu, Japan.

Schwalme, K., W. C. Mackay, and D. Lindner. 1985. Suitability of vertical slot and Denil fishways for passing north-temperate, nonsalmonid fish. Canadian Journal of Fisheries and Aquatic Sciences 42:1815–1822.

Silva, E. I. L., and R. W. Davies. 1986. Movements of some indigenous riverine fish in Sri Lanka. Hydrobiologia 137:265–270.

Stead, D. G. 1914. Notes on fisheries and fishery investigations in New South Wales during 1913. Pages 103–115 *in* Report on the fisheries of New South Wales, 1910-1913. Report to Legislative Assembly, New South Wales. New South Wales Printing Office, Sydney.

Talbot, G. B. 1952. Passage of shad at the Bonneville fishways. U.S. Fish and Wildlife Service Special Scientific Report Fisheries 94.

Travade, F. 1990. Monitoring techniques for fish passes recently used in France. Pages 119–126 *in* S. Komura, editor. Proceedings of the international symposium on fishways '90 in Gifu. Publications Committee of the International Symposium on Fishways '90, Gifu, Japan.

Vennard, J. K., and R. L. Street. 1982. Elementary fluid mechanics, 6th edition. Wiley, New York.

Welcomme, R. L. 1989. Floodplain fisheries management. Pages 209–233 *in* J. A. Gore and G. E. Petts, editors. Alternatives in regulated river management. CRC Press, Boca Raton, Florida.

White, D. K., and B. J. Pennino. 1980. Connecticut River fishways: model studies. Journal of the Hydraulics Division, ASCE 7(198):1219–1233.

Zhili, G., L. Qinhao, and A. Keming. 1990. Layout and performance of the Yangtang fishway. Pages 283–287 *in* S. Komura, editor. Proceedings of the international symposium on fishways '95 in Gifu. Publications Committee of the International Symposium on Fishways '95, Gifu, Japan.

Andrew W. H. Turnpenny and Julie K. Everard

Can Cavitation Injure Fish?

13

Scientific literature identifies several possible mechanisms (e.g., pressure flux, hydraulic shear, cavitation) through which hydraulic anomalies might lead to fish injuries. These mechanisms are of interest regarding the passage of fish through turbines and pumps; among them, cavitation is undoubtedly the least well understood. It is often downplayed by engineers who take pride in designing hydraulic machinery to eliminate equipment damage due to cavitation. Biologists may argue, on the other hand, that fish are more delicate than turbo-machinery and therefore more susceptible to cavitation effects. In this study, the material properties of fish tissues were examined in relation to possible cavitation effects. A spark gap apparatus was devised to generate single cavitation bubbles in proximity to the head and body surfaces of two species of fish (herring *Clupea harengus* and sole *Solea solea*). High-speed photography was used to form images of the development and collapse of the cavitation bubbles thus formed, and the fish were afterward examined for signs of injury. Experimental controls used 2 mm brass plate in place of the fish. The findings demonstrate that cavitation bubbles can collapse onto the surface of a fish's body but do not necessarily cause injury. The risk of injury must be related to both the probability of a fish coming into contact or close proximity to a vapor cavity and to the energy level associated with the cavity. It is unlikely that in the apparatus used the high energy levels associated with cavitation in a hydroelectric turbine were reached.

Introduction

There have been numerous studies of fish injuries resulting from passage through hydroelectric turbines. General reviews are provided, for example, by Eicher et al. (1987), Davies (1988), and Solomon (1988), while Cada and Coutant (1997) and Turnpenny (1998) considered experimental evidence relating to the specific causes of different injury types. These authors have ascribed observed fish injuries to a number of causes, including rapid pressure change, hydraulic shear, blade strike, and cavitation. Of these possible contributory causes, cavitation is the least well understood. It is the purpose of this paper to consider the fundamental question, "Can cavitation injure fish?"

Cavitation is the process of formation of gas bubbles in a liquid caused by a localized reduction in pressure to a point at or below vapor pressure (i.e., local boiling). The reduction in pressure may result from either a decrease in the ambient pressure, for example due to the lower hydrodynamic head on the downstream side of a turbine, or an increase in local velocity. Once formed, cavities may remain as vapor pockets attached to the blade or travel with the flow to regions of higher pressure, where they then collapse.

Since cavitation gives rise to noise and vibration and can materially damage turbo-machinery, turbine design and operating regimes are carefully engineered to avoid cavitation. Nevertheless, cavitation may occur if a turbine is operated away from its design conditions, a situation most likely to arise in tidal power schemes, where head- and tailwater levels may vary over a wide range (Solomon 1988). In an axial-flow turbine, cavitation typically occurs in the high-velocity regions along the leading edge of the runner or at the blade tip, where water leaks from the high pressure to low pressure side (tip leakage cavitation) (E. Goede and J. Pestalozzi, 1986, Sulzer-Escher Wyss, Ltd., Zurich, Switzerland, unpublished data). Dadswell et al. (1986) suggested that considerable cavitation occurred on the Annapolis tidal power scheme (Nova Scotia) at low tidal levels because of the shallow turbine setting.

The potential of cavitation as a source of fish damage is frequently mentioned in the literature but principally in relation to high-head schemes. In general, belief in the importance of cavitation as a source of fish damage seems to be based on the experience of pitting damage to turbines resulting from cavitation. Lucas (1962) pointed out that "while elimination of dangerous cavitation is known to be an objective of turbine designers…, complete safety for the fish requires reduction in cavitation beyond that which is safe for modern turbine. Obviously a fish cannot safely withstand the same forces as the turbine runner and casing materials." It should also be recognized that the turbines of that era were more prone to cavitation than in modern well-designed axial flow turbines where the hydraulics are optimized and the number of blades is few.

The mechanism for fish damage by cavitation is not clear. From early experiments, Muir (1959) developed the hypothesis that mortality among fish passing through high-head turbines is caused mainly by cavitation. His experiments involved generating a vapor cavity around the fish in an apparatus used for water hammer experiments. The pressure around the fish was reduced momentarily to vapor pressure, resulting in high mortality of Pacific salmon *Oncorhynchus* spp. fingerlings. Although vapor cavities large enough to accommodate a small fish might be generated in an operating turbine, it is difficult to envisage how a fish would enter such a cavity. It is doubtful, therefore, whether his findings are germane, except perhaps under the most extreme conditions of cavitation.

A more likely possibility, also recognized by Lucas (1962), is that fish could be exposed to damage from cavitation bubble collapse when small bubbles become detached from a vapor pocket and enter the flow. From investigations at the Annapolis tidal power scheme, Dadswell et al. (1986) cited "severe hemorrhages" and "pulping of body tissues" as evidence of cavitation damage, caused by "explosive release of vapor pockets."

The behavior of cavitation bubbles during collapse merits detailed consideration. Hammitt (1980) reviewed this subject. A bubble collapsing in midwater, away from any surface, will tend to collapse symmetrically, its energy density increasing as bubble volume shrinks, resulting in the emanation of a spherical shock wave at the final instant of collapse. Under this circumstance, the viscous force is no longer symmetrical, and water is pulled in preferentially from the side away from the surface (distal side), causing the bubble to flatten and collapse onto the surface. In some cases, the distal side will accelerate through the center of bubble to form a microjet, which impinges onto the surface at high velocity (Figure 1). Microjet impact has commonly been held responsible for the destructive action of collapsing cavitation bubbles on solid materials, though Fujikawa and Akamatsu (1980, Kyoto University, Kyoto, Japan, unpublished data) have demonstrated that it is the shock wave emanating at the instant of rebound which causes damage.

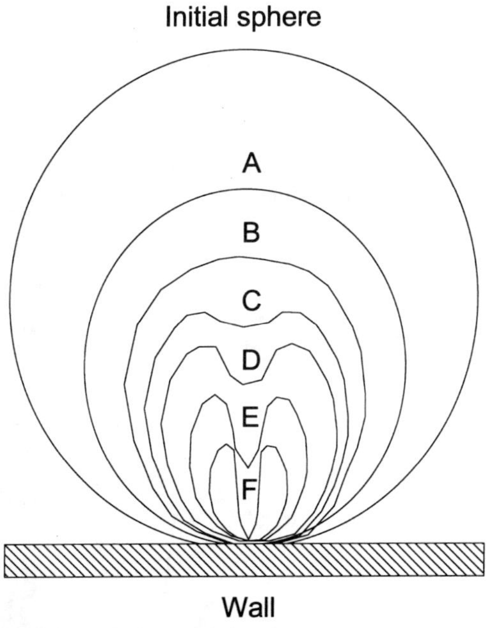

Figure 1. Cavitation bubble collapse onto a solid wall (after Hamitt 1980). As the bubble collapses (sequence A—>F), the upper surface inverts to form a microjet which, at point F, impinges upon the wall surface.

Bubble collapse behavior adjacent to a free surface (e.g., an air–water interface) is different, being influenced by the lower viscous forces on the proximal side of the bubble, adjacent to the free surface. In this case, the bubble collapses in the distal direction, moving away from the free surface. According to Hammitt (1980), this type of behavior may also arise where the surface is not free but elastomeric (e.g., rubber). In considering the significance of cavitation bubble collapse in the vicinity of a fish, we were prompted to question whether a fish would offer a surface more akin to an elastomeric or free surface rather than a solid surface. If so, a collapsing bubble would move away from the fish rather than toward it, thereby reducing the risk of damage.

To investigate this hypothesis, we conducted a simple experiment to compare cavitation bubble collapse behavior adjacent to a fish with that adjacent to a solid surface.

Apparatus and Methods

Individual cavitation bubbles were created using a custom-built underwater spark generator. The device produced an 8 joule spark across a 0.5 mm gap. In distilled water this resulted in a maximum bubble size of 8–10 mm, reaching peak expansion within 1400μs. The general arrangement used to hold specimens adjacent to the spark gap is shown in Figure 2.

The optical and electronic control system used to visualize and photograph cavitation bubbles is shown in Figure 3. The whole system was set up within a photographic studio where ambient light could be regulated. During experiments, the spark generator assembly was held immersed in distilled water contained within an optical-grade glass tank of dimension 75 × 190 × 75 mm. An image of the bubble was produced in

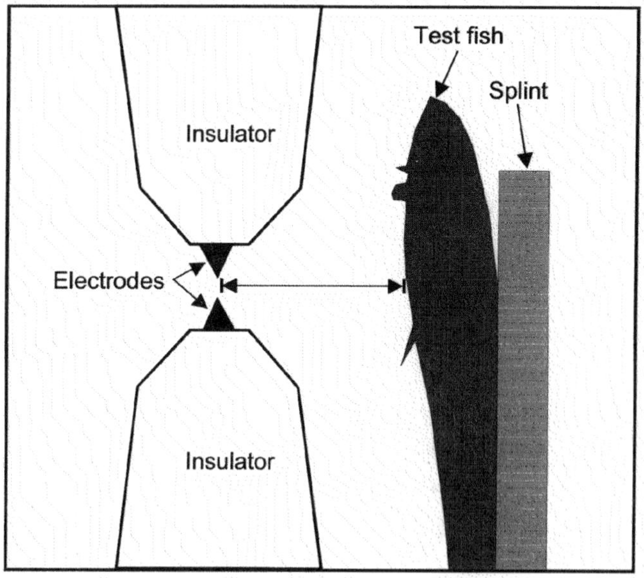

Figure 2. Diagram showing the positioning of the specimen relative to the spark gap formed by the electrodes. The double-headed arrow indicates where the distance between the specimen and the electrodes was measured. The spark gap was 0.5 mm.

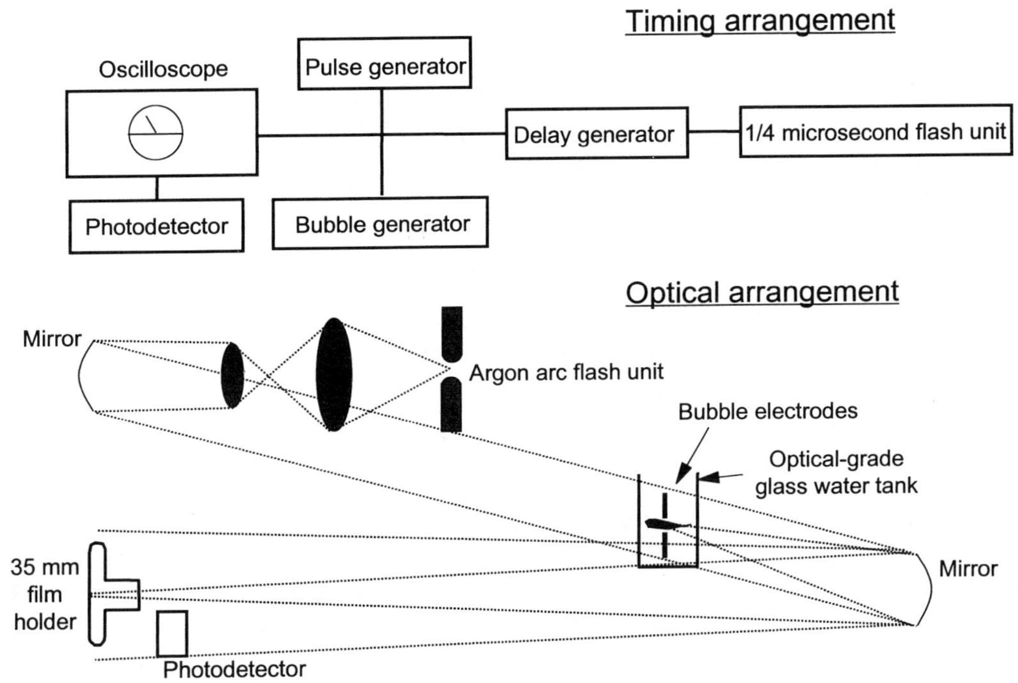

Figure 3. Experimental set-up: optical and timing arrangements.

silhouette using an argon arc light source with a 0.25 μs flash duration. This was focused onto a static 35 mm camera back. By means of the pulse generator and electronic time delay generator, which controlled the relative timing of the argon flash and spark generation, it was possible to take single-shot photographs at preselected intervals from the instant of spark initiation. The time taken to reach different stages of bubble development varied stochastically, with slow expansion of the bubble (1000–1400μs) followed by very rapid collapse (<100μs). As it was the collapse behavior that was of interest, shots were taken at various intervals between 1000 and 1300μs.

The size of fish that could be exposed to bubble collapse in this apparatus was limited by the free space within the electrode assembly. Trials were confined to two species, the sole *Solea solea* as age-0 juveniles of 32–42 mm standard length, and the herring *Clupea harengus,* of which a single 55 mm specimen was used. The fish were used fresh, within 5 min of immersion in a lethal concentration of MS-222 (1:5000). It was necessary to maintain a replicable geometry between the fish surface and the adjacent spark gap. This was achieved by mounting the specimen on a wooden splint, glued to the underside of the head of the fish using cyanoacrylate adhesive. The splint was then held in an adjustable clamp and its position relative to the electrodes adjusted to give the desired gap. This could be determined by temporarily placing a sheet of white paper behind the optical tank, onto which the silhouette of the electrodes and fish was projected. The gap was standardized using a caliper measuring gauge.

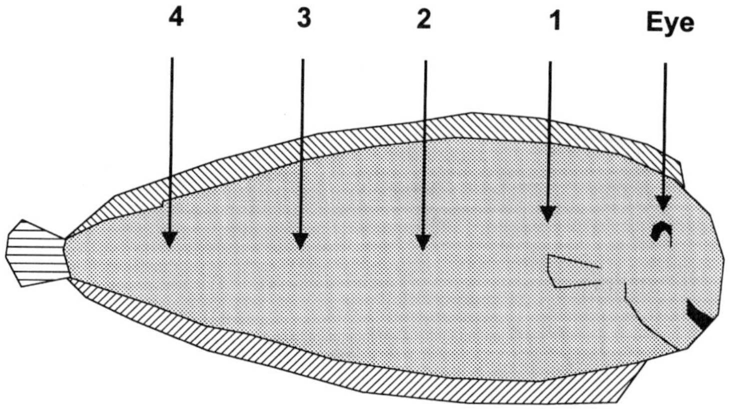

Figure 4. Positions along the body surface of the fish used for study of bubble collapse.

Individual fish were exposed to a series of five successive bubble implosions positioned at intervals along the head and body as shown in Figure 4. Each fish used was photographed before and after the series of exposures to record any damage existing before, or arising from bubble collapse. The fish were also examined under a binocular microscope to identify any superficial tissue damage, such as scale loss or pulping of the tissue.

For comparison with the fish trials, a piece of brass plate, measuring 19 × 19 × 2 mm, was used as a control specimen. Bubble collapse behavior adjacent to this plate was used to represent the control condition.

Results

Brass plate controls

An image of a bubble collapsing adjacent to the brass plate, taken at 1300μs after spark inception, is shown in Figure 5. The highly idealized bubble collapse behavior of Figure 1 is not fully evident in this photograph; nevertheless, overall asymmetry of bubble collapse is clear, with the implo-

Figure 5. Bubble collapse and impingement onto a brass plate, 1300μs after inception: the brass plate was placed 7 mm away from the spark gap.

sion being directed toward the metal surface. The brass plate experiment was repeated ten times, with the distance of the brass plate from the spark gap being set at either 7 mm or 10 mm. In all but one case, the bubble collapsed toward the brass plate (Table 1). Figure 6 (a and b) shows two stages of bubble collapse toward the brass plate observed in photographs similar to that shown in Figure 5.

Table 1. Experimental details and results from bubble collapse experiments.

Specimen	Specimen number	Body position struck by collapsing bubble	Time delay between generation and collapse of bubble μs	Distance from electrode mm	Bubble collapse direction (toward or away from specimen)
Brass plate	1	Not applicable	1300	7.0	Toward
	2	Not applicable	1300	7.0	Toward
	3	Not applicable	1300	7.0	Toward
	4	Not applicable	1300	7.0	Toward
	5	Not applicable	1300	7.0	Toward
	6	Not applicable	1300	10	Toward
	7	Not applicable	1300	10	Away
	8	Not applicable	1300	10	Toward
	9	Not applicable	1300	10	Toward
	10	Not applicable	1300	10	Toward
Sole	1	Eye	1300	6.8	Toward
		1	1300	6.8	Toward
		2	1300	6.8	Toward
		3	1300	6.8	Toward
		4	1300	6.8	Toward
Sole	2	Eye	1200	7.2	Toward
		1	1200	7.2	Toward
		2	1200	7.2	Toward
		3	1200	7.2	Toward
		4	1200	7.2	Toward
Sole	3	Eye	1100	7.2	Toward
		1	1100	7.2	Toward
		2	1100	7.2	Toward
		3	1100	7.2	Toward
		4	1100	7.2	Toward
Sole	4	Eye	1000	7.2	Toward
		1	1000	7.2	Toward
		2	1000	7.2	Toward
		3	1000	7.2	Toward
		4	1000	7.2	Toward
Sole	5	Eye	1300	4.4	Toward
		1	1300	4.4	Toward
		2	1300	4.4	Toward
		3	1300	4.4	Toward
		4	1300	4.4	Toward
Sole	6	Eye	1300	10	Away
		1	1300	10	Toward
		2	1300	10	Toward
		3	1300	10	Symmetrical
		4	1300	10	Toward
Herring	7	Eye	1300	7.2	Toward
		1	1300	7.2	Toward
		2	1300	7.2	Toward
		3	1300	7.2	Toward
		4	1300	7.2	Toward

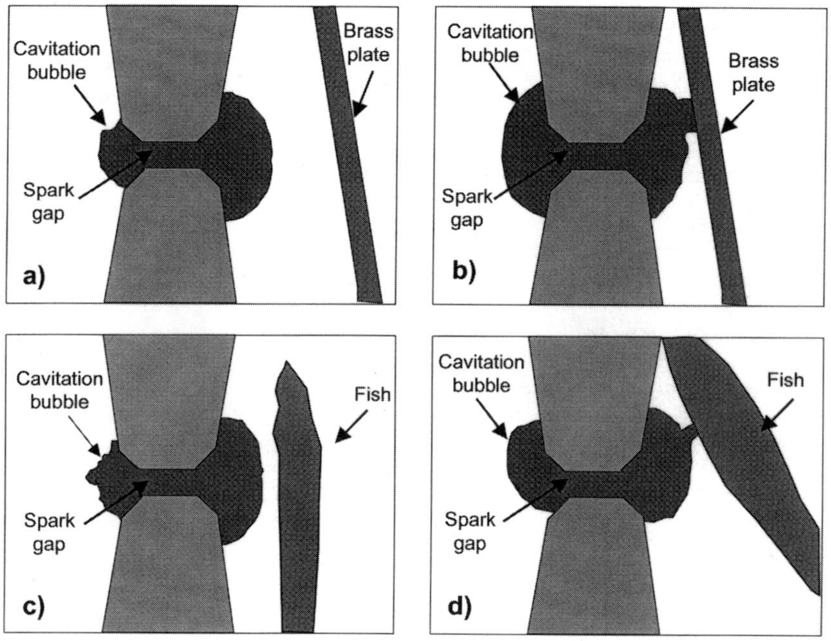

Figure 6. Collapse of cavitation bubbles onto surfaces. The upper two images show the asymmetrical initial collapse of the bubble (a), followed by microjet impingement (b) onto a brass plate. The lower images show the same collapse process when the brass plate is replaced by a fish (sole). The images were traced from actual photographs similar to that shown in Figure 5.

Fish experiments

Once the system had been adjusted and found to give good results with the brass plate, observations were made on seven fish specimens: six sole and one herring. This yielded 35 separate observations, the results of which and associated experimental conditions are summarized in Table 1. Bubble collapse was, in all but two cases (6%), directed toward the body surface of the fish. The exceptions both occurred with Specimen No. 6, in one case producing a symmetrical bubble and in the other a collapse directed away from the fish. This may have been owing to the fact that specimen no. 6 was placed at a greater distance from the spark gap (10 mm rather than 4.4–7.2 mm), thereby exerting less influence on bubble collapse. However, as the brass plate "controls" showed, some variance in implosion behavior can be expected, with no obvious cause. The pattern observed with herring (Specimen No. 7) was in all respects similar to that for sole. Figure 6 (c and d) gives examples of bubble collapse behavior toward fish, showing essentially similar behavior to that observed with the brass plate.

No evidence of tissue damage was found on any fish as a result of bubble implosion. Due to the highly consistent nature of the bubble collapse behavior, experiments were not continued beyond the series summarized here.

Discussion

The results of the single-bubble tests effectively disprove the hypothesis proposed in our introduction and demonstrate that the surface of a fish's body is analogous to the brass plate rather than the free surface model. It must be assumed that cavitation bubbles collapsing onto a fish's body could potentially damage fish, just as they may damage turbo-machinery, though, as Lucas (1962) points out, the damage threshold for fish tissues must be considerably lower. The fact that no damage to the fish resulted from bubble collapse in our experiments should not be interpreted as evidence that fish are safe from cavitation damage during turbine passage. Energy levels in the latter case must be vastly higher to account for the pitting that can be observed on turbine runners that have been allowed to cavitate (e.g., Goede and Pestalozzi, unpublished data). The question of cavitation's significance as a cause of fish damage during passage through a low-head turbine therefore reverts to one of statistics (i.e., the probability that a fish will come close enough to a vapor cavity to suffer damage), as well as of the energy level involved.

References

Cada, G., and C. C. Coutant. 1997. Development of biological criteria for the design of advanced hydropower turbines. U.S. Department of Energy, Report DOE/ID-10578, Idaho Operations Office, Idaho Falls.

Dadswell, M. J., R. A. Rulifson, and G. R. Daborn. 1986. Potential impact of large tidal power developments in the upper Bay of Fundy on fisheries resource of the Northwest Atlantic. Fisheries 11(4):26–35.

Davies, J. K. 1988. A review of information relating to fish passage through turbines: implications to tidal power schemes. Journal of Fish Biology 33(Supplement A):111–126.

Eicher, G. J., C. J. Campbell, R. E. Craven, and M. A. West. 1987. Turbine related mortality: review and evaluation of studies. Electric Power Research Institute, Report EPRI AP-5480, Palo Alto, California.

Hammitt, F. G. 1980. Cavitation and multiphase flow phenomena. McGraw Hill, New York.

Lucas, K. C. 1962. The mortality to fish passing through hydraulic turbines as related to cavitation and performance characteristics, pressure change, negative pressure and other factors. Pages 307–335 *in* Proceedings of the International Association for Hydraulic Research. Canada Department of Fisheries and Oceans, Vancouver.

Muir, J. F. 1959. Passage of young fish through turbines. Proceedings of the American Society of Civil Engineers, Journal of the Power Division 85(PO1):23–46.

Solomon, D. J. 1988. Fish passage through tidal energy barrages. Contractors Report Energy Technology Support Unit (ETSU) TID 4056, ETSU, Harwell, Oxon, UK.

Turnpenny, A. W. H. 1998. Mechanisms of fish damage in low-head turbines—an experimental appraisal. Pages 300–314 *in* M. Jungwirth, S. Schmutz, and S. Weiss, editors. Fish migration and fish bypasses. Fishing News Books, Blackwell Scientific Publications, Oxford.

Index